SCOTTISH ENGINEERING
THE MACHINE MAKERS

JAMES L WOOD

NMS Publishing Limited

Published by NMS Publishing Limited,
Royal Museum, Chambers Street, Edinburgh EH1 1JF

© James L Wood and NMS Publishing Limited 2000

Series editor: Iseabail Macleod

ISBN 1-901663-30-2

Other titles available in this series:

Building Railways	*Going to Church*	*Scots in Sickness and Health*
Farming	*Going to School*	*Scottish Bicycles and Tricycles*
Feeding Scotland	*Going on Holiday*	*Shipbuilding*
Fishing and Whaling	*Going to Bed*	*Spinning and Weaving*
Leaving Scotland	*Making Cars*	*Sporting Scotland*
Getting Married in Scotland	*Scotland's Inland Waterways*	

Forthcoming titles:

    Going to the Pictures   *Scottish Music Hall, Variety and Pantomime*

Other titles by the author include: *Shipbuilding* and *Building Railways*

British Library Cataloguing in Publication Data
A catalogue record of this book
is available from the British Library.

ISBN 1 901663 30 2

Internal design layout by NMS Publishing Limited.
Cover design by Mark Blackadder.
Printed and bound in the United Kingdom by Bell and Bain Limited, Glasgow.

Contents

Acknowledgements

The author gratefully acknowledges assistance received from the following sources:

John Crompton and Alison Morrison-Low, National Museums of Scotland; John Edwards, Aberdeen Maritime Museum; Geoff Hayes; the late Jimmie Houston; Stuart Liddle and Jim McCulloch, Hewlett-Packard Ltd; Dan Mackay; Willie Newlands, Hunslet-Barclay Ltd; Murdoch Nicolson, Mitchell Library, Glasgow; Frank Strange, Road Locomotive Society; Emrys Inker, Weir Pumps Ltd; Andrew Wood.

Illustrations: *pages* 10, 18, 20, 31, 33, 40, 41, 42, 47, 48, 49, 56, 67, 71, 73, 79, 95, 99 (Trustees of the National Museums of Scotland); 12 (Historic Scotland); 14 (Geoff Hayes); 17 (Barrhead Museum); 25, 53 (Dan Mackay Collection); 30 (Road Locomotive Society); 46, 55, 63 (*left*), 96 (Mitchell Library, Glasgow City Libraries and Archives); 50, 51, 52, 63 (*above*), 65, 77, 83 (James L Wood); 88, 89 (Hunslet-Barclay Ltd); 91 (Hewlett-Packard Ltd); 97 (Weir Pumps Ltd).

Introduction

Scottish engineering is a big subject with many facets, and it is inevitable that in a work of this length some topics will have to be omitted or given only a passing reference. The main focus will be on the development of the mechanical engineering industries in the nineteenth and twentieth centuries, with the addition of electrical engineering in later years. The emphasis will be on the origins and growth of what is regarded as the 'traditional' Scottish engineering industry, of which little now remains. Its decline will also be discussed although this is perhaps too recent, and too much a matter of current political concern, to be dealt with at great length. Civil engineering, that is the construction of canals, railways, roads, bridges, docks and other public works, will not in general be included. However, something will be said about iron and steel bridges and other aspects of structural engineering. Although civil engineering has been covered to a limited extent in the 'Scotland's Past in Action' series, in *Building Railways* and *Waterways,* it is a major subject which really requires its own book, or books.

Other titles already published within the series include *Spinning and Weaving* and *Shipbuilding.* These topics are linked to *Building Railways* by the engineering industry which made the spinning and weaving machinery, the engines which powered the ships and the locomotives and other equipment for the railways. They were the major sectors of the Scottish engineering industry although their relative importance changed with time. But many other types of machinery were made over the years and these, though less well-known, collectively formed a significant part of the industry's output. The range of products included mining and quarrying equipment, machinery for the sugar industry, printing and paper-making machinery, range-finders for battleships and periscopes for submarines, driers and ironing machines for laundries, woodworking machinery, stone-working machinery, machines for bakers and confectioners, lawn-mowers and agricultural implements, and even beer engines for public houses.

Scottish society of today has largely been shaped by 250 years of industrial development. The engineers played a major role in this process and for most of the nineteenth century the Scottish engineering industries were world leaders. It has been said, not unreasonably, that Scotland's national dress should be the boiler suit, rather than the kilt! The twentieth century has seen great changes. Modern Scottish industry now has a high proportion of factories owned by companies based outwith Scotland. Much of the engineering design and development work, together with strategic planning and decision-making, therefore takes place outside Scotland and frequently outside the United Kingdom. It is not a great exaggeration to say that there is no longer a Scottish engineering industry. Instead there is an international engineering industry, part of which happens to operate in Scotland. The problems which can arise in this situation are obvious enough. In bad times the branch factories are likely to be the first to close and there is little which can be done in Scotland to influence closure decisions.

Engineering is frequently misunderstood and confused with science. One dictionary definition is the 'application of science for control and use of power, esp. in roads and other works of public utility (civil engineering), machines (mechanical engineering), electrical apparatus (electrical engineering)'. There is, however, much more to engineering than the application of science. The successful manned flights to the moon from 1969 onwards were sometimes described as great scientific achievements. However, one or two engineers did offer the suggestion, with tongue in cheek perhaps, that the main scientific contribution was the calculation of the position of the moon at the relevant time! The creation and operation of the Saturn rocket and the lunar landing vehicle that made it possible to put men on the moon and return them to earth was the work of engineers. There are, of course, many aspects of engineering which have a much more immediate impact on daily life than space flight. We are dependent on engineers for the necessities of urban life, a supply of clean water (and disposal of the dirty water), electricity and gas, and transport and communications of all types. These tend to be taken for granted, except when they fail for whatever reason. More highly valued by many people, if not so essential, are things such as domestic appliances and entertainment systems. Engineers are concerned with the creation of all of these. Engineering is the process of conception, design and production of useful machines and structures. It combines creative art and science. Faced with the problem of designing and building a machine to a deadline and within a budget, an engineer will use what scientific information is available and relevant. Beyond that a possible solution has to come from creative thinking, experience and a knowledge of what has worked (or not worked) in the past. Engineering design almost always involves a step, sometimes large sometimes small, into the unknown. The productions of engineers are as much expressions of human creative ability as paintings or musical compositions, and in their own way they are works of art. They can have a special fascination too, and not just for schoolboy trainspotters! The illustrations have been chosen to show something of the range of the output of Scottish engineering firms over the years. There was not space to mention all of the important and interesting products and manufacturers in the main text and in selecting illustrations the opportunity has been taken to repair some of the omissions.

In any field of human endeavour it is almost inevitable that some achieve a prominence which tends to overshadow the efforts of others. The history of engineering offers numerous examples of people who have made significant contributions but who are now unknown to the general public and to most present-day engineers. Of course something must be said about the work of the major figures, but in addition an attempt has been made to rescue a few of the unknowns from undeserved obscurity.

1
Industrialisation and the
Birth of the Engineering Industry

During the second half of the eighteenth century there began what has become known as the Industrial Revolution. This is usually seen as the replacement of small-scale manufacturing processes, carried on by hand as cottage industries, by power-operated machines in factories. If used without qualification the term tends to give the impression that the scale of operations before the Industrial Revolution was insignificant and also exaggerates both the speed and the completeness of the change. The process of industrialisation started with the textile industry, when the first water-powered spinning mills were built. Not only was the spinning technology new, but there was also a new fibre being processed, imported cotton.

For centuries the production of cloth, both wool and linen, was by far the most important manufacturing industry in Scotland. While production was on a domestic scale, the cloth was by no means all for domestic use and the total output was large. Much of it was sold through merchants in the towns and the organisation of this domestic industry was a major operation. The Woollen Manufactory of Glasgow, which was set up in 1699 and organised the textile production of 1,400 people, was exceptional in scale although it was by no means the only company of this type. By the union of the Scottish and English Parliaments in 1707 linen production had become the most important section of the textile industry. The Board of Trustees for Improving Fisheries and Manufactures, set up in 1727, made great efforts to improve quality at all stages, from the growing of the flax to the finishing of the woven cloth. In the middle of the eighteenth century the textile industry, headed by linen, was a huge and complex organisation of merchants, 'putters-out' and middlemen bringing together the work of those involved in all stages of the process from growing or importing the flax, through spinning and weaving to bleaching and finishing. The range of products was wide, from cheap cloth for slaves in the colonies to copies of expensive continental and Indian fabrics for the fashion-conscious. Although cloth production mostly involved hand processes, in both the woollen and linen industries there were stages in production which were carried out in specialised 'factories' using machinery driven by water wheels. These factories can be regarded as forerunners of the power-driven spinning mills of the Industrial Revolution.

The machines which made power spinning possible were the inventions of James Hargeaves (*circa* 1764), Richard Arkwright (1769) and Samuel Crompton (1779), all from the north of England. There is scope for discussion about whose contribution was the most important technically, but there is no doubt that Arkwright was the most effective in exploiting his invention. His first factory was in Nottingham, modest in size and worked

Water-powered beetling machines were used to improve the texture and finish of linen or jute cloth and yarn by prolonged pounding with wooden mallets. This yarn-beetling machine was installed at Baluniefield near Dundee, in the first half of the nineteenth century. It is now displayed in the Museum of Scotland.

by horses. The second, built at Cromford, Derbyshire, in 1771, was powered by a waterwheel and it was this which became the prototype for the many cotton spinning mills built wherever a suitable water supply was available. Within a few years spinning mills were being built in Scotland, the first at Penicuik in 1778. The Scottish mills were largely financed by Scots, although Arkwright had an interest in one or two, notably New Lanark, where the first of several mills was in production by 1785. Much of the initial technical expertise also came from the earlier mills in England. Among the early water-powered cotton spinning mills in Scotland was that at Catrine, in the parish of Sorn, Ayrshire. In the Old Statistical Account, written late in the eighteenth century, it is described as

> … entirely a new creation, and owes its existence to the flourishing state of the cotton manufacture in Great Britain. In the year 1787, Mr. Alexander of Ballochmyle, the proprietor of the village, in partnership with the patriotic Mr Dale of Glasgow, built a cotton twist-mill … with a fall of water … of 46 feet. A jeanie factory and a corn mill are drove by the same fall …. Three hundred and one persons, old and young, are just now employed, in carding, roving, and in spinning, with an overseer and two clerks: Clock-makers, smiths, millwrights, and other mechanics, amount to 15 more …

[The 'jeanie' is the spinning jenny, invented by James Hargreaves.]

Claud Alexander had been an official of the East India Company and it was from this that he made the money for investment in Catrine mill. The Company held the monopoly of trade with India and in practice governed the country. Much of the capital which went into new industrial ventures at this time came from people who, like Alexander, had done well in the colonies. A large water wheel, built in 1826 by Hewes & Wren of Manchester for Grandholm mill, Old Machar near Aberdeen, is now in the collection of the National Museums of Scotland. About 1898 this was moved to Woodside mill, Aberdeen, which had been built as a cotton-spinning mill two years before Catrine and was latterly associated with the paper industry. The mill was finally closed in 1966 and the wheel was acquired by what was then the Royal Scottish Museum.

The mechanisation of spinning was easier than the next process, weaving. Although Edmund Cartwright patented the first power loom in 1785 it was far from perfect. It was well into the nineteenth century before the technical problems involved in weaving by power were completely solved. Even so cotton goods, woven on hand looms using machine-spun yarn, were much cheaper than linen or woollen articles made entirely by hand and the cotton industry expanded dramatically. A large proportion of the finished product was exported, much of it to the areas from which the raw cotton had come.

These early factories were complex entities. They were driven by large water wheels, with a power transmission system of gears and belts to the individual machines. The number of sites suitable for water wheels of the size necessary was limited and the later mills were powered by steam engines. The actual cotton-processing machinery was rather like large-scale clockwork made of wood and metal, and repeated thousands of times. The buildings themselves were on a bigger than usual scale and posed new problems in construction.

Water wheel of 150 hp (112 kw) built in 1826 by Hewes & Wren of Manchester and used in the Aberdeen area. The wheel was 25 ft 2 inches (7.7 m) diameter and 21 feet (6.4 m) wide. With these large wheels the drive was taken through a gear ring fixed directly to the rim, rather than from the axle as was usual with small water wheels. From about 1898 it worked at Woodside mill, a former cotton-spinning mill which was latterly associated with the paper industry. When this closed in 1966 the wheel was acquired by the Royal Scottish Museum and was displayed there for many years.

As indicated above, water-powered textile machinery had been used for some processes well before the first cotton spinning mill was built. Water-powered machinery was also used in mining to drain pits and wind coal, and in the iron industry. The steam engine invented by Thomas Newcomen was in use for mine pumping well before the first cotton mills were built. His first engine was erected in 1712 at a colliery near Dudley Castle, in the English Midlands. Several engines of this type were erected at Scottish pits around 1720, the first probably at Tranent. Before 1787, when the Catrine cotton mill was created, twenty-five or more pumping engines had been built in Scotland. Even earlier were the numerous water-powered grain mills. The creation of these machines was an engineering job, although the people involved were not then described as engineers but by their various trades: 'Clock-makers, smiths, millwrights, and other mechanics.' Of these the millwrights were to be particularly important in the development of the engineering industry in Scotland in the first half of the nineteenth century.

The early spinning mills required substantially more power than other types of water-powered mills. They therefore had to be located in country areas, because only there were there sites with sufficient water available. Suitable sites were few in number. Steam-driven mills could be sited anywhere provided there was a supply of coal, easy access to raw materials and a supply of labour. In 1796 there were thirty-nine cotton mills in Scotland, all in rural areas and driven by water wheels. By 1812 there were 120 cotton mills, with most of the increase being accounted for by steam-driven mills in and around Glasgow.

As the textile mills, and especially the steam-powered mills, grew in number so the opportunities expanded for businesses to provide engineering services. An important figure in this new industry was millwright and blacksmith James Cook, who was born in Fife in 1764. He moved to Glasgow in 1788 where he set up his own workshop, initially carrying out the same kind of work which he had done in Fife. It would not be stretching things too far to say that this was the first mechanical engineering firm in Scotland and it marked the beginning of the engineering industry as we know it. With the expiry in 1800 of James Watt's patent for the steam engine with a separate condenser, Cook began to build steam engines. These included engines for the new cotton-spinning mills then being established in the city. He also pioneered the manufacture of steam-driven sugar mills,

Glenruthven mill, Auchterarder, was a cotton-weaving mill set up in 1877. The original looms, belt-driven from overhead shafts, were made by the Anderston Foundry Company, Glasgow. Some of these were in use until production ceased about 1980. The mill was powered by a steam engine and coal-fired boiler right to the end. (Geoff Hayes)

making use of the long-established financial links between the west of Scotland and the sugar plantations in the West Indies. This was a significant development and one which was to be of great importance in later years. The growth of Cook's business led him to build a new works in 1805 in Tradeston, Glasgow. This, the first purpose-built engineering factory in Scotland, was regarded as a bit of a gamble and it was therefore designed to be readily convertible to dwelling houses if the venture did not succeed. Happily this never became necessary. An early job for the new works was the building in 1806 of a steam engine for the Clayslap flour mill, Partick, the first flour mill in the Glasgow area to be driven by steam. Where Cook led others followed and within a few years there were several such general engineering firms in being. They had gained the knowledge and experience which would enable them to participate in what was to be the 'sunrise' industry of the time, steam navigation.

2
Steam Power at Sea

It is likely that the first attempt to use a steam engine to propel a boat was made in France in the 1770s, but there is little information available about the vessel itself or about what, if anything, was achieved. When the time was right, both technically and commercially, the concentration of expertise gained through the use of steam engines for industrial purposes meant that the breakthrough came in the west of Scotland. The steamship and the industrial developments which flowed from it were to change the basis of the Scottish economy. A great many people were involved in this process, some well-known, many more whose contributions have been forgotten. In what follows there is space to deal only with the main threads of the story and with a few of those concerned.

The Scottish connection with steamships begins with the work of William Symington (1764-1831). Although his fame rests principally on his important contribution to the development of steam navigation, this was only a small part of his working life. He began his engineering career at the Wanlockhead lead mines, where he was involved in the construction of machinery, including steam pumping engines. He left Wanlockhead in 1791 and by 1808 he had designed and built some thirty engines. Most of these were for Scottish collieries, but in 1788, while still at Wanlockhead, he produced his first engine for a boat. This was for an experimental vessel which was tried, with limited success, in 1788 on Dalswinton Loch, Dumfriesshire. In the following year further trials were made with a larger engine, this time on the Forth & Clyde Canal. This work was funded by Patrick Miller, owner of the Dalswinton estate, who had made his fortune as a banker in Edinburgh. Far more important technically were Symington's two vessels constructed in 1801 and 1802 with the support of the Forth & Clyde Canal Company, which was interested in the possible use of steam-boats to tow barges on the Canal. The second vessel was the well-known *Charlotte Dundas*, which was fitted with a single-cylinder horizontal engine driving the single paddle wheel near the stern. This horizontal engine was, by comparison with the complicated devices used previously by Symington himself and others, a model of simplicity and it became one of the most common forms in use throughout the subsequent history of the steam engine. The *Charlotte Dundas* worked well during trials in 1803 but the canal company had fears, not wholly unjustified, that the wash might damage the banks and did not take things any further. The vessel lay on the canal at Falkirk for several years and was seen by a number of people with an interest in steam navigation, among them an American, Robert Fulton. On his return to America Fulton built the *Clermont*, in which was installed a steam engine bought from Boulton & Watt of Birmingham and a London-made boiler. With this ship he began the world's first commercial service by steamboat, on the Hudson River in 1807.

For the development of the engineering industry in Scotland, an important landmark was Bell's *Comet* of 1812, the first steamship to operate a commercial service in Europe. Henry Bell (1767-1839) came from a family of millwrights, a highly skilled trade encompassing building, metalwork, woodwork and requiring a knowledge of the behaviour of flowing water (or hydraulics) in order to design waterwheels and the channels which fed them. He trained as a stonemason and millwright, then made it his business to obtain some experience of shipbuilding at Bo'ness. For eighteen months from 1788 he worked in London with fellow Scot, the renowned civil engineer John Rennie, when the latter was engaged in the Albion Flour Mills, a large mill driven by Boulton & Watt steam engines. Henry Bell then practised as a civil engineer, builder and property developer in Glasgow. In this role he built the Baths Inn, Helensburgh, in 1806-7. He disposed of the property in 1810 but continued to live there and manage it, although there must be a strong suspicion that his wife did most of the managing while he was busy with numerous other ploys!

Bell wanted to increase trade at the Baths Inn, Helensburgh, by making it easier for his patrons to travel from Glasgow. This was the commercial incentive for the building of the *Comet*. The technical elements of the vessel, its wooden hull, engine and boiler, were all routine products of local industry by this time. The hull of the *Comet* was built at Port Glasgow by the firm of John Wood & Company which was to go on to build many steamboats in the next forty years until iron began to replace wood. Two Glasgow engineers, John Robertson (1782-1868) and David Napier (1790-1870), supplied the engine and boiler respectively. Robertson came from Neilston, where his father was foreman in a cotton spinning mill. He had been apprenticed at the age of fourteen to a local spinning-wheel maker, which must have been a dying trade with the development of spinning by power. On finishing his apprenticeship he went to work as a turner (or lathe operator) at the Stanley cotton mills, near Perth, where his father had been before moving to Neilston. In 1810 John Robertson started his own engineering business in Glasgow. Among other things he worked on steam heating installations for mills and on improvements to textile machinery. The engine of the *Comet* was a variation on the normal beam engine, with the beam mounted low down and pivoted at one end rather than in the middle. David Napier was the son of John Napier, who had a blacksmith's business and foundry, initially in Dumbarton and then from 1802 in Glasgow. His son in due course learned his trade with him, but seems not to have been formally apprenticed. By 1810 he was in complete charge of his father's business, possibly because the latter was ailing (he died in 1813). Making the boiler for Bell caused some problems because it had an internal

John Robertson (1782-1868) and the engine installed in Henry Bell's steamboat Comet *in 1812. When the ship was wrecked in 1820 the engine was salvaged and used for many years to drive machinery ashore. Recognising its historical importance, Robert Napier & Sons purchased the old engine and presented it to the Museum of Patents, South Kensington, London, the forerunner of the Science Museum.*

David Napier (1790-1869), maker of the boiler for Bell's Comet. *While he was a notable marine engineer, his contribution to the development of steamship services was possibly even more important. By establishing services between Scotland and Ireland, he demonstrated that it was practical to use steamships in the open sea as well as in river estuaries.*

flue, a complication to which Napier was not accustomed. Napier also made the castings required for the engine. Both Robertson and Napier had had past business dealings with Henry Bell and were happy to accept his orders for the work. Sadly their trust proved misplaced. Bell had serious financial problems and neither was ever paid in full. Where Bell led, others rapidly followed. Even before the *Comet* was in service the next steamer was under construction and by 1816 twenty-six ships had been built. Here was a brand new market for the engineer and at least six firms tried their hand in this period.

Of the marine engineers already mentioned, the most significant was to be David Napier, the maker of the boiler for the *Comet*. As early as 1814 he had seen the potential of the steamship and begun the construction of a new works at Camlachie, Glasgow. There he built his first marine engine in 1816, for the *Marion*. This was also the first of a number of ventures into shipowning. He stayed at Camlachie until 1821 when, 'finding that steam navigation was becoming an important business, I purchased lands at Lancefield, on the banks of the Clyde, as being more suitable for the purpose than Camlachie'. In 1836, by which time he had built engines for over forty vessels, he decided to move to London. Although he developed and patented several new ideas while there, as a business venture the move was a failure.

Important though David Napier was, the leading light in marine-engine building in the first half of the nineteenth century was his cousin, Robert Napier. He too was born in Dumbarton where his father, James, worked with David's father, John. Robert Napier served a formal apprenticeship with his father, after which he moved for a time to Edinburgh and then to Glasgow. There he worked with William Lang, who made equipment for the textile industry, before setting up his own business in 1815. He built his first marine engine for the paddle steamer *Leven* in 1823 and that engine still survives at Dumbarton. Although conventional beam engines were sometimes used on board ship, especially in the United States, the weight of the high-mounted beam could cause a problem with stability. British marine engineers much preferred a rearranged form of the beam engine, in which the beam was in two parts mounted one on either side of the engine and low down in the ship. This was known as the 'side-lever' engine and the *Leven* engine is of this type.

If Robert Napier was the leading light in the industry, then it is fair to say that a good deal of the fuel was provided by David Elder, who joined Napier in 1821 as manager. Elder was born in Kinross. His father was a millwright and David Elder was also trained as a millwright. Both Napier and Elder therefore came to Glasgow from the same sort of background of millwrights' and blacksmiths' work in a small town. The scope of their activities grew with the expanding industry and Robert Napier followed David Napier into the Camlachie works and then to Lancefield when the latter moved on. The firm steadily built up a reputation for first-class work and much of the credit for this must go to David Elder. In addition to making sure that the engines were solidly built and self-contained, so that they were not dependent on the ship's hull for rigidity, he developed many new machine tools and manufacturing techniques to speed production. David Elder may be undervalued by history, but Robert Napier did not make that mistake. In 1835 he entered into a new agreement with Elder, under which he was paid an annual salary of £250 plus 7/6d (37½p) per Nominal Horse Power of the engines which Napier contracted to build. Nominal Horse Power (or NHP) was a calculated figure for engine power output based only on the physical dimensions and an assumed steam pressure. As the design of engines improved over the years the actual power output became much greater than the nominal power, although the NHP value could still be useful as an indication of the size of an engine and the amount of work involved in its construction. In the context of David Elder's remuneration the important thing is that the figure for NHP could be large, totalling perhaps 2,000 or more in a year. This would add £750 to his basic salary, bringing the total to £1,000. This is equivalent to £100,000 or more at the present day.

After fifteen successful years as a marine-engine builder with a growing

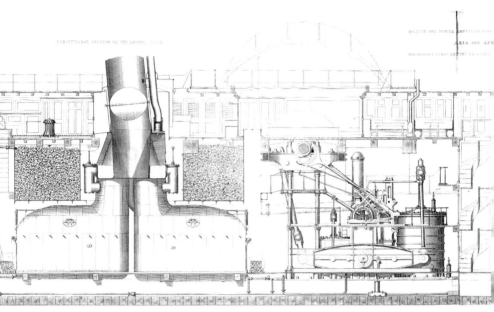

Arrangement of the engines and boilers of the paddle steamers Asia *and* Africa, *1850.*
Samuel Cunard began his transatlantic steamship service with the Britannia *and three*
sister ships built in 1840-41. These were followed by a series of ever larger and more
powerful versions of the same basic design. The Asia *and* Africa *were both built of wood*
by Robert Steele & Company of Greenock, and Robert Napier of Glasgow built the side-
lever engines and the box boilers. This superb drawing was executed by David Kirkaldy, a
Dundee man who served his apprenticeship with Robert Napier and became his chief
draughtsman and calculator. He subsequently became renowned in the engineering world
as one of the pioneers of the systematic testing of the strength of materials.

reputation among his customers, Napier received in 1838 the first of many
orders for engines for ships of the Royal Navy. Almost all of these had
previously come from London firms. He was pushed onto the world stage
by the arrival of transatlantic steamship services and his links with Samuel
Cunard. The early steamers were used on coastal traffic and short sea
crossings, but the dream of a transatlantic service had formed early on.
Others had started services across the Atlantic some two years before
Cunard. This gave him, and Napier, the opportunity to learn from other
peoples' mistakes. Significantly, Cunard had been awarded a mail
contract, the thing which was to prove indispensable to almost every
successful transatlantic shipping line. The first four Cunard ships, with

which the service was inaugurated in 1840, were wooden paddle steamers. They were ordered from Napier as main contractor in 1839 and he subcontracted the hulls to various Clyde shipyards. They had twin cylinder side-lever engines of 420 NHP and about 740 actual horsepower (550 kw), with cylinders 72.5 inches (1.84 m) diameter x 82 inches (2.1 m) stroke. It was about this time too that Napier began to think about becoming a shipbuilder as well as an engineer. In 1841 he acquired ground in Govan and there he established a yard to build ships of iron. It was the engineers rather than the shipbuilders who had taken the lead in development of the steamship and it was natural that they should take up shipbuilding in iron.

A steamship was not necessarily any more comfortable than a sailing ship, but the passages were shorter and more predictable. Charles Dickens wrote a graphic account in *American Notes* of a voyage in January 1842 on board the *Britannia*, one of the four original vessels which Cunard ordered from Robert Napier in 1839. The weather was bad throughout and the passage from Liverpool to Boston took eighteen days. This was perhaps four days longer than it would have taken in good weather, but a sailing ship in those conditions might well have taken anything up to six weeks. By the end of the century the fastest mail steamers were taking about six days for the crossing and even the slowest passenger liners would cross in ten days or so. For all passengers, but in particular for the many emigrants, which even the best ships were by then carrying in large numbers, steam power was a boon and a blessing beyond price.

Because Napier's firm dominated the industry on the Clyde in the early decades of steam, it is not surprising that many of the second generation of marine engineers learned the business there, as apprentices, journeymen, draughtsmen and managers. So well did they learn that, by the 1860s, the firms established by those trained at Napier's were providing the technical leadership and taking business away from the 'old school'. In the following pages many ex-Napier men will be featured.

To begin with it was enough that ships could proceed from A to B with reasonable reliability. Soon the reliability was taken for granted and for merchant vessels fuel consumption became a matter of great concern. Steam was viable on short routes with dense passenger traffic, or long routes such as the Atlantic where again there were many passengers and, more important, governments were prepared to support the service by means of payments for the conveyance of mail. In 1862 Robert Napier built the *Scotia* for Cunard, their last paddle steamer. In essence, it was little different from the ships with which the service was started in 1840. The *Scotia* was larger and had a hull of iron instead of wood, but the machinery was just a bigger version of the same low pressure side-lever engines. It was not just a matter of running costs, although this was important. The quantity of coal which had to be carried left little room for paying cargo. Moreover, there were few facilities for coaling and this severely limited the use of steam on long routes. Brunel's *Great Eastern*, launched at Millwall on the Thames in 1858, had to be the enormous size it was because it was designed to carry the coal needed for a voyage to Ceylon (now Sri Lanka) and back. The Admiralty has often been accused of lacking interest in steam propulsion, but this is not really justified. They were aware of the usefulness of steam, but short range and the scarcity of coaling stations meant that sails remained essential on ships which had a worldwide operational role. The time had come to go beyond making the steamship work reliably. It was necessary to make it cheaper to operate and thereby widen its usefulness.

The way forward lay in the reduction of fuel consumption and in this process Scottish engineers played an important part. A major step was adoption of the compound engine, in which the steam is expanded successively in two cylinders of different sizes. Most of the credit for this development is usually given to John Elder, son of David Elder, Napier's manager, but there is more to the story than that. Three other men, Charles Randolph, John L K Jamieson and William McNaught, also had significant roles to play in the introduction of the compound engine. John Elder served his apprenticeship at Napier's under his father and then, after a spell in England, he returned to Napier's around 1848 as head draughtsman. Charles Randolph was another Napier apprentice. Following his apprenticeship he went to Manchester to work for two firms of millwrights. He returned to Glasgow in 1834 and set up his own company. Different partners worked with him over the years until he was joined by John Elder in 1852, when the firm became Randolph, Elder & Company. Until then the firm's work was the equipping of mills with engines and power transmission systems, throughout Scotland but particularly in Glasgow and Dundee. With the arrival of John Elder there was a shift of emphasis to the expanding field of marine engineering and, some years later, shipbuilding also. John L K Jamieson served his apprenticeship at the Canal Basin Foundry, Port Dundas, Glasgow. He then had a variety of experience in Ireland and England, including a spell in the Royal Navy as a sea-going engineer, before becoming superintendent engineer for the Pacific Steam Navigation Company. William McNaught was yet another who became a Napier apprentice, 'ere he had quite completed his fourteenth year, working diligently at his trade by day, and attending science classes at the Andersonian University in the evenings'. (The Andersonian University was the forerunner of the University of Strathclyde.) After a period in charge of a mill in India, which did not agree with his health, he came home and joined the consulting engineering business of his father, John McNaught.

John Elder was no doubt the driving force behind the introduction of the compound marine engine, but his patent of 1853, *Improvements in Propelling Vessels*, was actually taken out in the names of Charles Randolph and John Elder, in that order. This claimed 'the application to the propelling of steam vessels, by means of the screw or other submerged propeller, of engines having double cylinders, for the purpose of using the steam expansively', in other words compound engines. The idea of the compound engine was far from new and it had been successfully applied to a number of river paddle steamers based in the Netherlands from 1829 onwards. Gerard Maurits Roentgen, who was responsible for these, had obtained in 1834, via an agent Ernst Wolff, a British patent for the compound engine. If, as seems likely, Randolph and Elder knew of this patent, they would have known that it would be difficult to defend any general patent covering the compound marine engine. Their patent was therefore confined to vessels driven by screw propellers. On land compound engines had been tried, with some success, in the first decade of the nineteenth century, but it was the form patented in 1845 by William McNaught which became important. He was consulted by Glasgow cotton manufacturers about increasing the power of their existing old beam engines. These were single-cylinder machines with the cylinder and crankshaft at opposite ends of the engines. McNaught conceived the idea of adding another smaller cylinder halfway between the bearing on which the beam pivoted and the crankshaft. This made the engine into a compound and at the same time, because the piston thrust from the new cylinder acted in the opposite direction to that of the original cylinder, the stresses in the beam were reduced. The original boilers, which were likely to be well-worn, could be replaced while the engine was being altered by new ones working at a higher pressure. The result was a more economical engine of greater power for much less than the cost of a completely new power plant. The McNaught system could of course be applied also to a brand new engine.

Randolph, Elder & Company's first compound engine was installed in the *Brandon* in 1854 and in the following year the Pacific Steam Navigation Company, who operated on the west coast of South America, where coal was expensive, placed an order for compound engines for their new ship *Valparaiso*. It was about this time too that John Jamieson became superintendent engineer of the company and in that capacity he was much in Randolph, Elder's works while the engines were under construction. Now Jamieson records that while he was at the Canal Basin Foundry he 'had charge of the erection of the first pair of altered compound engines on McNaught's plan in 1848 at Messrs Johnstone and Galbraith's Spinning Mills in Glasgow'. Here we have Randolph and Jamieson, both with experience as millwrights and therefore well aware of the benefits of compounding as demonstrated by the McNaught engine, one the principal partner in the engine builder and the other occupying the senior engineering position with the customer. By 1866 Randolph, Elder & Company had built forty-eight sets of compound engines, many for the Pacific Steam Navigation Company. Not only were they installed in new ships, but the older vessels were sent home in turn to be re-engined. A reduction in coal consumption approaching fifty per cent was claimed. Jamieson left the Pacific Steam Navigation Company in 1866 and joined Randolph, Elder as general engineering manager. Two years later he became a partner when Charles Randolph retired. Following John Elder's sudden death in 1869 at the early age of forty-five, a new partnership was set up as John Elder & Company, thus capitalising on his great reputation. One of the partners in the new company was John Jamieson. He remained a partner until 1879 and died in 1883, highly regarded by his contemporaries but a man rarely mentioned in the history of marine engineering.

With the economic benefits of the compound engine well established, the idea suggested itself that there might be further benefit in expanding the steam in three stages instead of two. Although Elder had patented a form of triple-expansion engine in 1862, it was to be many years before they were much used. The engineer most closely associated with the triple-

Triple expansion marine engines at various stages of construction in the Govan works of McKie & Baxter. Founded about 1896, they were specialist builders of small and medium-sized engines and boilers. (Dan Mackay Collection)

expansion engine was Alexander C Kirk. Yet another of those who spent his early years at Napier's, he went south to work for one of the leading marine engineering firms in London, Maudslay, Sons & Field. When he returned to Scotland in 1865 it was not to marine engineering but to the shale-oil industry at Bathgate. Following the death of John Elder in 1869, Kirk became manager of the engineering department of the new partnership, John Elder & Company. In 1874 a triple-expansion engine designed by Kirk was installed in the *Propontis*, belonging to a Liverpool shipowner, W H Dixon. The new engine worked well, but unfortunately the ship was

25

fitted with a second innovation in the form of high-pressure boilers of a novel design. These suffered two serious explosions while the ship was at sea and, not surprisingly, shipowners were in no rush to give the triple-expansion engine a further trial. Then in 1881 the Aberdeen firm of George Thomson & Company had the *Aberdeen* built for them by Robert Napier & Sons. The triple-expansion machinery was again designed by A C Kirk, but the boilers were of a conventional and well-tried type. Kirk was by this time the senior partner in Napier's. Robert Napier had died in 1876 and there was no one in the family interested in carrying on the business. It was therefore sold to a new partnership headed by Kirk. The *Aberdeen* proved to be a complete success and was followed by many thousands of ships with similar machinery. Ships powered by triple-expansion steam engines were still being built by Scottish shipyards in the 1950s. Generalisations can be dangerous, but contemporary opinion suggested that a good compound engine would use just over half the coal used by a simple expansion engine; a triple would save a further fifteen to twenty per cent. Under the right conditions an additional small saving might be achieved by expanding the steam in four stages, or quadruple-expansion, but this was not often worthwhile. The compound engine's reduced fuel consumption brought the operating cost of steam ships below that of sailing ships on most routes. As the use of the triple-expansion engine became widespread the role of the ocean-going sailing ship was reduced to vanishing point.

To obtain the maximum benefit from compound engines, the steam pressure had to be raised above that commonly used with 'simple' engines in which the steam was expanded in a single stage. Expansion in three or four stages required further increases in steam pressure. The improvements in engine efficiency were therefore dependent on the availability of boilers suitable for working at the higher steam pressures. The most efficient shape for a boiler to enable it to resist the pressure of the steam would be a sphere, but this is difficult to make and the cylindrical form became the usual one. However, early boilers were rectangular in cross-section. Pressures were very low, and the shape was easy to make and more convenient for fitting into a ship. Standards of workmanship could leave much to be desired. In that context the comments of Peter Denny, of the associated Dumbarton firms of William Denny & Brothers (shipbuilders) and Denny & Company (engineers), are interesting; he was talking to the Institution of Naval Architects in 1891 about the practice of some forty years earlier:

Seven pounds of pressure if you could get it, if not as much as you could. I remember the boilers of that time, with splendid stalactites of salt on their fronts and elsewhere, and the comforting assurance of the foreman boilermaker that they would soon 'tak up', which they did in a way, often with the assistance of horse manure. It was about this time, possibly a few years later, that a well-known Clyde engineer, still alive, told me that he had been asked to arbitrate in a question about a boiler which had been complained about as defective. He said he went to inspect it with a view to helping a brother chip out of a difficulty if at all possible. His remark to me as to the result was, 'Haud in watter? It wadna haud in sma' tatties'. ['Hold in water? It wouldn't hold in small potatoes.']

It is not surprising that increases in boiler pressure were treated cautiously! To cure boiler leaks the horse manure was of course applied internally. His remark about stalactites formed on leaky boilers highlights the fact that marine boilers then used salt water and this in itself placed a restriction on the maximum useable steam pressure.

The early compound engines used steam at about twenty-five pounds per square inch (1.7 atmospheres). At this pressure the box boiler could still be made sufficiently strong by judicious use of stay bars connecting the pairs of flat sides. As working pressures were increased to fifty pounds per square inch (3.4 atmospheres) and beyond, in the search for higher efficiency, box boilers had to be abandoned in favour of cylindrical boilers. There were various forms of these but the one which was most widely used originated on the Clyde. It was built worldwide but wherever it was made it was known as a Scotch boiler. Because the Admiralty tended to be conservative in the matter of steam pressures, box boilers were retained for naval vessels for a good number of years after they had been given up in merchant ships. Curiously it is not certain who was the originator of the Scotch boiler, although it has been suggested that it might have been the Glasgow engineer James Howden, about 1860.

The *Propontis*, as mentioned above, was fitted with boilers of an unusual design when the triple-expansion engine was installed in 1874. This decision was prompted by the desire to use a higher steam pressure than that which was possible with the Scotch boilers of the day. Water-tube boilers working at 150 pounds per square inch (10.2 atmospheres) were therefore installed. These were to the design of J M Rowan, a Glasgow engineer, and T R Horton of Birmingham. The first of this type had been made as far back as 1858, but although Rowan and Horton had shown commendable determination over many years they were never able to make their boiler sufficiently reliable for it to come into general use. Those on board the *Propontis* were still regarded as experimental and following the explosions already mentioned they were replaced by Scotch boilers working at a lower pressure.

The Scotch boiler is of the fire-tube type with the hot gases from the furnace passing through numerous small tubes immersed in the water-filled shell of the boiler. In a water-tube boiler the water is inside the tubes and the hot gases are on the outside. The boiler contains much less water and it is therefore lighter in weight and more easily designed to withstand high pressures, because there is no large water-filled shell which is subjected to the full working pressure. The water-tube boiler is, unfortunately, one of these devices which seems so simple and obvious but in practice, as Rowan, Horton and many others discovered, it proved to be full of surprises. In particular it is much more fussy about the quality of the water than the Scotch boiler. Most of the sediment from the water in a Scotch boiler accumulates on the bottom of the boiler shell, where it does little harm as this is the coolest part. With a water-tube boiler, deposits will be inside the tubes which are subjected to the full heat of the furnace. These deposits impede the transfer of heat to the water, causing over-heated and burst tubes, and this appears to have been the problem with the boilers of the *Propontis*. By the time the *Aberdeen* was built in 1881, mild steel was replacing wrought iron. With this stronger and more consistent material, Scotch boilers could be built to work at the higher pressures needed to get the full benefit offered by the triple-expansion engine. It was not until the late 1890s that water-tube boilers became widely used at sea, initially in naval vessels. One of the most successful boilers of this type was that developed in America by Babcock & Wilcox and patented in 1867. Until about 1890 they were used only on land. The Babcock boiler owed much of its success to the fact that the tubes were of relatively large diameter and straight, which made cleaning easier than it was with some rival designs. In the mid-1880s the firm began making boilers in Scotland, initially in part of the then new Singer factory at Clydebank and subsequently at a new works established at Renfrew in 1897.

The Glasgow engineer James Howden, and his firm James Howden & Company, were instrumental in bringing into general use two important developments in marine boiler practice, forced draught and oil-firing. Forcing air into the furnace, instead of relying on the natural draught created by the funnel, had been tried in the very early years of steam navigation but had no lasting success. Howden began work on forced draught in the 1860s, but it was 1884 before the system with which his name is associated was first used. The key feature was that the air for combustion was pre-heated by the hot gas from the furnaces before it was sent up the funnel. Howden's forced draught system brought a significant increase in efficiency and was used in a great number of merchant ships. Oil-firing of boilers was tried to a limited extent in the late nineteenth century and interest grew in the early twentieth century, especially for naval vessels. In association with the Tyneside firm, the Wallsend Slipway and Engineering Company, James Howden & Company worked on the problems involved and developed the Wallsend-Howden oil fuel system which became widely used.

After the end of World War I, oil was increasingly used as fuel on merchant ships and especially the large passenger liners. As these ships were refitted after their war service, conversion of the boilers to oil-firing was usually included in the work done. With coal-fired boilers a large crew was needed in the engineering department. For example, before conversion the Cunard liner *Mauretania* had an engine room complement of 366. Of these, 192 were firemen and 120 were trimmers who brought the coal from the bunkers to the boilers. After conversion to oil only seventy-nine men were needed in the engineering department. For the shipping companies the main attraction of oil as fuel was therefore a reduction in operating costs. Working in the stokehold of a large coal-fired ship was one of the dirtiest and physically most demanding jobs that there was. Conversion greatly improved working conditions for the few men now required, but it is likely that the men who were unemployed as a result would have mixed feelings.

By far the most powerful steam engines built for any purpose were the marine engines. The largest built on the Clyde developed about 15,000 hp (11,200 kw) and cylinder diameter could be nine feet (2.75 m) or more. By the 1890s there were perhaps thirty or so firms building marine engines, two thirds of them shipbuilders who had their own engine works. However only a handful of firms, the builders of large liners and warships such as Fairfield Shipbuilding & Engineering Company (Govan), J & G Thomson (Clydebank) and Scott's Shipbuilding & Engineering Company (Greenock), could build engines in the region of 10,000 hp (7,500kw) and over. At the other end of the range anybody could build the small engines, although it might not pay the large firms to do so. Once the basic design of the

Two Scotch boilers each being hauled by two steam traction engines of an unusual type designed by R W Thomson, a consulting engineer in Edinburgh, and known as 'road steamers'. Those Clyde shipbuilders who did not have cranes suitable for installing heavy items after a ship had been launched, used those belonging to the Clyde Navigation Trust. This involved towing the ships alongside the cranes and then bringing the boilers and engine parts to the cranes by road.

compound and triple-expansion engine had become established these were needed in large numbers, in a range of sizes but all basically similar. No great inventiveness was required and consequently this was a market which new firms could easily enter. Although the leading shipbuilders had their own engine works, there were many others who did not and relied instead on specialist engineering firms. Typical of these was Dunsmuir & Jackson, of Govan. This was founded in 1878 by Hugh Dunsmuir and William Jackson, both former Napier apprentices who had worked for other firms in various parts of the country before setting up on their own. In 1888 it was reported that they employed 400-500 'hands' and devoted 'special attention to marine engines on the now widely prevalent triple expansion principle, and produce a splendid class of engines of this kind ranging as high as three thousand horse-power'. The shipyards with their own engine works also made boilers, as did many of the specialised engine-builders such as Dunsmuir & Jackson. There were, however, firms which came to be primarily boiler-makers. Among these was Anderson & Lyall of Govan, founded in the 1870s, a time of rapid expansion for the Clyde shipbuilding and marine engineering industries.

In addition to the engine-builders and boiler-makers there were many makers of various fittings and pieces of auxiliary equipment. As ships grew in size and complexity there was more and more of this kind of equipment. It was usually cheaper and more convenient for firms to buy in such items rather than to make them. For example boiler fittings, such as water level and steam pressure gauges, safety valves and other items, were almost always bought in from one of the many specialists. G & J Weir were particularly noted for the many types of pump required on board ship, especially those used for feeding water into boilers. This firm originated in Liverpool but moved to Glasgow in 1873. Not all makers of equipment for the shipbuilding and marine engineering industries were in

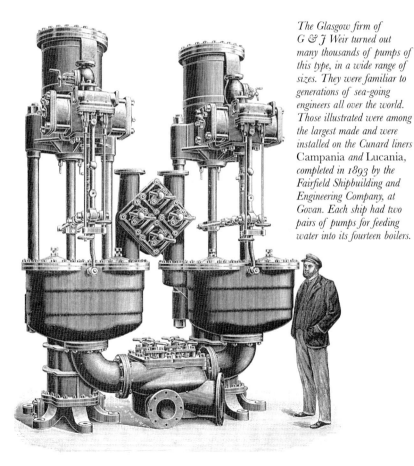

The Glasgow firm of G & J Weir turned out many thousands of pumps of this type, in a wide range of sizes. They were familiar to generations of sea-going engineers all over the world. Those illustrated were among the largest made and were installed on the Cunard liners Campania *and* Lucania, *completed in 1893 by the Fairfield Shipbuilding and Engineering Company, at Govan. Each ship had two pairs of pumps for feeding water into its fourteen boilers.*

the west of Scotland. For example Brown Brothers was established in Edinburgh in 1861 as general and hydraulic engineers. Power steering systems were increasingly necessary for ever-bigger ships and a steam steering engine or 'steam tiller' was patented in 1888 by A Betts Brown. The firm also installed complete hydraulic systems on board ship. These could be designed to operate everything from the steering gear and cranes for cargo handling, to passenger lifts and hoists to take food from the galley to the dining saloon. Shortly before the outbreak of World War II, in conjunction with the Dumbarton shipbuilder William Denny & Brothers, a hydraulically operated fin stabiliser was developed to reduce rolling. Initially it was fitted to warships as an aid to accurate gunnery, but after the end of the war it quickly became a standard fitting on important passenger ships, including the *Queen Mary* and *Queen Elizabeth*.

An important innovation in marine propulsion, from which much was expected although it never fully lived up to these expectations, was the steam turbine. The first practical turbines appeared in the 1880s, notably that of Charles Parsons, patented in 1884. Within a few years numerous turbines were being used to drive generators for electric lighting on board ship and one ship so fitted was Clyde paddle-steamer, *Duchess of Hamilton*, built in 1890 by William Denny & Brothers. Ever since this initial contact between Parsons and Denny's, consideration had been given to the building of a turbine-powered ship. In 1894 Parsons built a successful small experimental turbine-driven vessel, the *Turbinia*, but no shipowner was willing to take the risk of placing the first commercial order. The Turbine Steamer Syndicate was therefore formed to finance and operate the world's first turbine-powered merchant ship, the Clyde steamer *King Edward*, completed by William Denny & Brothers in 1901. Denny's and the Parsons Marine Steam Turbine Company, Newcastle, provided the money, and Captain John Williamson, a well-established operator of steamer services on the Clyde, managed the ship. The speculators were handsomely rewarded and by 1906 Denny's had built seventeen turbine steamers. For the first few ships, including the *King Edward*, the turbines came from Parsons but very soon they were being built at Dumbarton by Denny & Company, the marine engineering firm associated with William Denny & Brothers. Where Denny's led others soon followed and the main Scottish firms entered into agreements with Parsons to build turbines. The Admiralty very soon realized the potential of the steam turbine for Royal

Model of the 6,500 hp (4,850 kw) compound diagonal engines installed by Denny &
Company in the paddle steamers Princesse Henriette *and* Princesse Josephine, 1888.
The ships were built at Dumbarton by William Denny & Brothers for the Belgian
government's Dover-Ostend service. This magnificent model, one-twelfth full-size, was made
by the Glasgow engineer and instrument maker, David Carlaw, for Denny's.

Navy warships, but they were careful to adopt a belt-and-braces approach and were keen to have an alternative design to the Parsons turbine available in Britain. They therefore encouraged John Brown & Company to obtain a license in 1908 to build the Curtis turbine at Clydebank, in addition to the Parsons type which there were already building by then. The Curtis was an American design, which offered the possibility of saving space and increasing efficiency. In the end it was the Curtis turbine which became prone to serious problems as it was developed for increased power outputs. From 1930 only Parsons turbines were installed in ships being built for the Royal Navy. It is worth noting that the manufacturing licenses were used by John Brown's as platforms for further development. Manufacturing under a licensing arrangement is of limited value if, as happened all too often, the licensee remained totally dependent technically on the licensor.

To begin with there was in some quarters the view that the turbine was but a passing fashion. There was even a suggestion that it just would not do at sea because it was delicate and the engineers could not get at it with a hammer! At the other extreme, the turbine came to be viewed by some as the ideal type of machinery for almost all ships. However, time was to show that while the turbine had important advantages it also had limitations. It was lighter in weight, took up less space and produced less vibration than reciprocating engines, the last mentioned being specially valuable in passenger ships. The turbine therefore became the natural choice for warships, large liners and fast cross-channel ships, where very high power outputs were required with minimum space and weight. For example the Cunard liner *Lusitania* built by John Brown in 1907 had turbines of 72,500 hp (54,100 kw), an output which could not have been achieved in the space available using reciprocating engines. One problem with the turbine was that its optimum rotational speed was much greater than that of a propeller. The early ships, including the *Lusitania*, had the two directly coupled so that the choice of speed was a compromise which was too low for the turbines and too high for the propellers. The answer was to connect the two by means of speed-reducing gears, but at that time no one had ever made gears capable of transmitting the power required. The problems associated with gearing took time to solve, but late in 1911 the Fairfield Shipbuilding and Engineering Company completed the first passenger steamers with geared turbines, the London & South Western Railway's cross-channel ships *Normannia* and *Hantonia*. An alternative to mechanical gearing was the use of electrical power transmission. The Glasgow-based firm of William Beardmore & Company took out a license for the Swedish-designed Ljungstrom steam turbine and electrical transmission system. There was only one installation, on the *Wulsty Castle*, completed in 1918 at Sunderland. This gave much trouble and Beardmore's did not pursue the idea any further. However, in America and to a limited extent in Britain, other firms using different machinery designs had more success with turbo-electric propulsion. Despite the efforts outlined above to widen the scope of the marine steam turbine, its range of application remained limited. It was never economical enough to displace the reciprocating engine in, for example, tramp steamers. In 1912 the diesel engine was first used, very successfully, in ocean-going ships. Since then the power output and efficiency of the diesel have increased to the point where it has replaced not only the reciprocating steam engine but also the steam turbine in all but a few very specialised applications, such as the nuclear submarine. The Scottish response to the diesel engine is outlined in Chapter 5.

3
The Specialists

The most important part of the engineering industry in nineteenth-century Scotland was marine engineering, as discussed in the previous chapter. But there were other specialisations within the industry. By the end of the century there were hundreds of firms, making a wide range of products. In what follows it has been possible to mention only a few of the firms and the most important of the products. As Scotland became more and more industrialised, the early engineers moved into new fields of activity. In time, provided they survived the financial hazards of a rapidly changing business world, these 'generalists' tended to become specialists in one or two areas. James Cook, as already noted, had started by making machinery for grain mills and cotton mills, and then expanded into marine engineering and sugar machinery. About 1830, after making around twenty marine engines, the decision was taken to concentrate almost entirely on the manufacture of sugar machinery, an industry which Cook had pioneered. In contrast, firms which were established in mid century and later were often specialists from the start.

Another Glasgow firm which came early to marine work was Murdoch & Cross which was founded in 1815 and turned out their first marine engine four years later. Better known by their later name, Murdoch, Aitken & Company, they were the first firm to build railway locomotives in Glasgow. Two were built in 1831 for the Monkland & Kirkintilloch Railway and from such modest beginnings there grew up another important part of the Scottish engineering industry, although Murdoch, Aitken themselves had little further involvement. Even in the middle of the nineteenth century, when a tendency towards specialisation was clearly discernable, a contemporary advertisement indicates that Murdoch, Aitken & Company were still prepared to make almost anything!

After ships the best known product of Scottish industry was the railway locomotive. Over the years some thirty firms made locomotives in Scotland but most of these built only a handful. The main centre of the industry was Glasgow, although significant numbers were produced in Kilmarnock and Leith, and firms in Dundee, Greenock and elsewhere built small numbers of locomotives. The oldest of the three main Glasgow builders was Neilson & Company, founded in 1836 as a general engineering firm under the name of Mitchell & Neilson; after several changes in the partnership it eventually became Neilson & Company in 1855, under the control of Walter Montgomerie Neilson. His father was James Beaumont Neilson, who had introduced the use of pre-heated air in the blast furnaces of ironworks. This innovation resulted in significant savings in fuel costs for the owners of the works, and considerable wealth for Neilson. It was he who had provided much of the finance for the firm right from its foundation by James Mitchell and William Neilson. The latter was a cousin of Walter Montgomerie Neilson, who did not himself become involved until 1838. Products included factory driving engines, for which they had a good reputation, marine engines, sugar machinery and plant for the mining and gas industries. The first locomotives were built in 1844, for the Garnkirk & Glasgow Railway, and such was the success of the venture into locomotive building that the decision was taken by 1852 to specialise in this field. For a few years the other products were continued alongside the locomotive work. The original works were in the Finnieston district of Glasgow, but in 1860-1 the firm moved a to new works at Springburn.

Soon after the move to Springburn Neilson's works manager and partner, Henry Dübs, left to set up his own firm of Dübs & Company in the Queen's Park district of Glasgow. Dübs had come from Germany in 1842 to work in the English locomotive-building industry. In 1857 he was dismissed from his job as works manager with the Manchester firm of Beyer, Peacock largely, it seems, because he was too easy-going. Nevertheless Neilson was prepared to take him on, partly because Dübs had contacts with the English railway companies from whom Neilson was looking for orders. To make room for Dübs it was necessary to get rid of his existing works manager, James Reid. Before long the relationship with Dübs had gone sour. When he left he took with him Neilson's chief draughtsman and other key staff. In addition, several of Neilson's customers transferred their loyalty to the new firm. The first locomotive was turned out in 1865, a year after construction of the factory started. Early in 1866 the firm pioneered the employment of women as tracers in the drawing office, a practice which quickly spread to other firms. Until then women had had no place in the engineering industry.

When Dübs left Neilson's, James Reid was brought back in 1863 as manager and partner under a ten-year agreement. Before his first spell with Neilson's, James Reid had worked with the Greenock firms of Caird & Company and Scott, Sinclair & Company. These were marine engineering and shipbuilding firms, but both had dabbled briefly in locomotive building. After leaving Neilson's he went to work for the Manchester firm of Sharp Stewart & Company, one of the leading firms in the industry, and this experience would have made him well-fitted to join Walter Neilson once again. However in the early 1870s, as the ten-year period was nearing its close, Neilson and Reid had a serious disagreement. While the details are not entirely clear the upshot was that it was Neilson who left his own firm. By 1876 Reid had total control of Neilson & Company. One cannot help wondering whether Reid had been quietly nursing his anger until he could have revenge for his earlier dismissal. In 1884 Walter Neilson tried to re-enter locomotive building in Glasgow and bought ground in Springburn, almost next door to his old company, to set up the Clyde Locomotive Works. The first locomotives were completed there in 1886, but despite Neilson's long connection with the industry orders proved hard to find. In 1888 the works was sold to Sharp, Stewart & Company who wanted a replacement for their old works in Manchester. Part of the deal was that Neilson should have a directorship for life in the firm, but he did not live long to benefit from this as he died in 1889. A significant latecomer to locomotive building in the Glasgow area was the industrial giant, William Beardmore & Company, who built them at their Dalmuir works, to the west of Glasgow, from 1920-30 as part of a post-war diversification programme.

There were a number of firms who specialised in locomotives for use at collieries, ironworks and other industrial sites. Of these by far the most important in Scotland and one of the most important in Britain was the Kilmarnock firm of Andrew Barclay, Sons & Company. The firm was established by Andrew Barclay in 1842 and soon the main business was supplying winding engines and other equipment to collieries. Building small locomotives for internal use at these collieries and other industrial sites was a natural development which began in 1860. Over the years they were built in many sizes for various industries and a few main line locomotives were also made. The National Museums of Scotland collection includes a diminutive two-foot (610 mm)-gauge example, which was used within Granton gasworks, Edinburgh. The only sizeable locomotive builder in the east of Scotland was begun in Leith as an offshoot of the Newcastle firm of R & W Hawthorn. It was set up in 1846 to assemble locomotives for Scottish railway companies from parts sent by sea from Newcastle. When the last link in the east-coast main railway line was finished in 1850, completed locomotives could be delivered by rail. The works in Leith was then sold to a new firm, Hawthorns & Company, who continued to build locomotives for many years and also made marine engines. On display in the Museum of Scotland there is the industrial locomotive *Ellesmere* which was built by Hawthorns in 1861 for use at a colliery in Lancashire.

The main Scottish railway companies, in addition to buying locomotives from specialist builders, also constructed some in their own workshops. The North British and Caledonian Railways had works in Glasgow at Cowlairs and St Rollox, close to the works of Neilson and Sharp Stewart in Springburn. With these four major works Springburn was at one time the most important locomotive building centre in Britain. The Glasgow & South Western Railway works were at Kilmarnock, the Highland Railway in Inverness, and the Great North of Scotland Railway at Inverurie near Aberdeen. These railway workshops built carriages and wagons as well as locomotives and there were also two important firms of carriage and wagon builders, R Y Pickering of Wishaw and Hurst, Nelson & Company in Motherwell. The latter firm built tramcars as well as railway rolling stock.

The building of stationary steam engines was carried on by many firms throughout Scotland. Engines were built in the cities, in virtually all large towns and some surprisingly small ones also. The National Museums of Scotland has an important collection of engines and something will be said here about the makers and various uses represented. A few of the Glasgow general engineering firms which made stationary engines have already been mentioned. The rapidly developing cotton-spinning industry was an important market for them, but by the middle of the nineteenth century the industry in Glasgow had reached a plateau and there can have been only a modest demand for new engines. Thereafter, while stationary engines were still built in large numbers in Glasgow and elsewhere in the west of Scotland, most were built as adjuncts to the main business of the various makers. For example, the Coatbridge firm, Lamberton & Company, a leading builder of rolling mills for steelworks, also made the engines which powered them. The manufacturers of sugar machinery built many engines to drive sugar-cane crushing mills and steam-driven vacuum pumps used in sugar refining. In the National Museums of Scotland collection there is a single-cylinder horizontal vacuum-pumping engine built in 1919 by the Harvey Engineering Company of Glasgow for a London sugar refinery. Two of leading engine builders in the east of Scotland are also represented: James Carmichael & Company, Dundee, and Douglas & Grant of Kirkcaldy. The Carmichael engine is a large horizontal compound of about 300 hp (220 kw), built in 1923 to power the Wemyss Sawmill, Leven. Founded in 1810 as J & C Carmichael, this firm was one of the oldest of several engineering and steam-engine building firms in Dundee. In 1833 they made two locomotives for the Dundee & Newtyle Railway but did not pursue this branch of the industry any further, building instead stationary engines, boilers and gearing for mills.

The author's interest in the engines of Douglas & Grant was kindled forty years ago when he first saw models of the Corliss-valve engines in the then Royal Scottish Museum. Both had been made in the Museum workshop, one representing an engine of the 1870s and the other showing the early twentieth-century development of the same basic design. A full-size engine of 60 hp (45 kw), dating from 1923, was subsequently acquired from a woollen mill in Alva, Clackmannanshire, and this is now displayed in the Museum of Scotland. The firm was established by Robert Douglas in 1846, initially in Cupar and from 1854 in Kirkcaldy. To begin with the usual wide range of engineering work was undertaken. Output included engines, boilers, waterwheels, machinery for the paper industry and even one of the few steam road rollers constructed in Scotland. Through contact with a Canadian-born engineer William Inglis, who was trained in Scotland and subsequently practiced as a consulting engineer in Edinburgh, Robert Douglas was introduced to the form of steam engine developed in 1849 by an American engineer, George Corliss. In 1863 Douglas built a Corliss engine to Inglis' design for a papermill on the Water of Leith, the first engine of this type to be produced in Britain. A technical description of the workings of the Corliss engine would be out of place here. Suffice to say it offered better speed regulation and possible economies in fuel by comparison to earlier engines. They became very popular for factory driving and other purposes, and their manufacture was taken up by virtually every major stationary engine builder in Britain. Over 200 Corliss engines were made by Robert Douglas or the later partnership, Douglas & Grant. Lewis Grant had been associated with Douglas at least since 1869, when a patent relating to Corliss engines was taken out in their joint names. He became a business partner in 1872 and son-in-law at some date unknown. The Corliss engines ranged in power output from 60 hp (45 kw) up to 2,500 hp (1,865 kw). The latter engine was built about 1888 for a cotton mill belonging to the Mazagon Spinning and Manufacturing Company, Bombay.

The willingness of Douglas & Grant to make use of good ideas from overseas did not end with the Corliss engine. In 1910 the firm obtained a license to manufacture engines to the design of the Belgian firm, Carels

A beam engine built by Douglas & Grant of Kirkcaldy about 1877, shown after the initial erection in the maker's works. This is an example of a 'house-built' engine, one which will be partly supported by the walls of the engine house when it is finally installed. It is a compound engine on the system patented by William McNaught in 1845, with the high pressure cylinder positioned between the centre of the moving beam and the crankshaft.

The steam engine was not only the source of power for the mill but, as here, it was often a show-piece. Important visitors would be taken into the highly-finished engine house to be impressed by the prosperity and stability of the firm. This compound engine of about 2,000 hp (1,490kw) was built about 1880 by Fullerton, Hodgart & Barclay of Paisley for one of the local mills.

Brothers, of Ghent, which was in turn derived from a nineteenth-century design by the Swiss firm of Sulzer Brothers, Winterthur. Curiously, the design of the Carmichael engine referred to above shows a pronounced continental influence. The transfer of the technology from Winterthur to Kirkcaldy, via Ghent, was by formal licensing agreements. How the final transfer from Kirkcaldy to Dundee was effected is not known, although one can always speculate!

The Kilmarnock firm of Andrew Barclay has already been mentioned in connection with locomotive building and as builders of winding engines and other machinery for the mining industry. Another Kilmarnock firm in the same line of business was Grant, Ritchie & Company, set up in 1876. The founders, Thomas Grant and William Ritchie, had both worked for Barclay's and it is not surprising that some of their products closely resembled those of their former employer. It has even been said that they were seen leaving Andrew Barclay's works one night with rolls

Steam winding engine installed at Lady Victoria colliery, Newtongrange, Midlothian in 1894. It was originally built by Grant, Ritchie & Company of Kilmarnock, but in the course of its long working life many parts were renewed by Andrew Barclay, Sons & Company, Kilmarnock. This is a twin cylinder simple engine, that is both cylinders are the same size. About seven tons of coal at a time was raised from a depth of 1,624 feet (500 m). The photograph was taken early in 1981 and shows winding engineman, Archie Smith. The colliery closed not long afterwards. It is now the home of the Scottish Mining Museum and the engine remains in its original location.

of drawings tucked under their arms! By far the best surviving example of a Scottish-built steam winding engine is that preserved by the Scottish Mining Museum at the former Lady Victoria Colliery, Newtongrange, Midlothian. This was built by Grant, Ritchie in 1894, but winding engines have a hard life and extensive repair work has been carried out by Barclay's over the years and it is probably now more of a Barclay engine than a Grant, Ritchie. There were firms who built both marine and stationary engines, especially in the early part of the nineteenth century. One firm making both types at the end of the nineteenth century and the beginning of the twentieth century was Fleming & Ferguson of Paisley. Established in 1885, they were also shipbuilders specialising in dredgers. Their engine works not only built machinery for their ships but made a sizeable number of engines for factory driving and for pumping water and sewage. They even managed to win orders for pumping engines from London County Council in competition with English firms who specialised in this type of machinery.

Although there were firms in Glasgow producing textile machinery as early as 1801, it was not unknown for mills to make machinery in their own workshops. For example, the cotton-spinning mill set up in the Anderston district of Glasgow about 1804 by Henry Houldsworth & Company had an associated machinery manufacturing organization which became a separate company, the Anderston Foundry Company, in 1823. Both spinning machinery and power looms were made to begin with, but in later years emphasis was on looms of many different types. In addition to making cloth from wool, cotton and the numerous synthetic materials now available, wire cloth is also produced. This has many uses in, for example, filtration equipment and in paper making. The Anderston Foundry Company was one of a number of Scottish makers of special looms for this work. An exception to the general rule that firms founded after the middle of the nineteenth century tended to specialise from the start was A F Craig & Company of Paisley, established in 1868. They made a wide range of products which included textile machinery of various types. The National Museums of Scotland has one of their Wilton carpet looms made about 1900. While the Glasgow cotton industry was stagnating in the second half of the nineteenth century, the Dundee jute industry was expanding rapidly and partially displacing the older flax industry. The manufacture of textile machinery for flax, jute and other fibres was a major industry in the area. After cloth of any kind (other than wire cloth) has been woven it requires 'finishing', that is bleaching and perhaps dyeing or printing, before it can be sold. There were several Scottish engineers engaged in the manufacture of the machinery used in these finishing processes. One of the most important was the Glasgow firm of Duncan Stewart & Company, set up in 1864 to make textile printing machines and other finishing equipment.

The manufacture of sewing machines in Scotland is of special interest. Although machines were being made in Glasgow as early as 1860 by R E Simpson & Company, the great bulk of the production was carried out in factories owned by American companies. The Singer Manufacturing Company began building sewing machines in Scotland in 1867 at Bridgeton, Glasgow. In 1884 they moved to the Kilbowie factory near Clydebank which, with its huge numbers of employees and special-purpose machine tools, was unlike any other works in Britain at that time. By the turn of the century production was some 13,000 machines a week and 7,000 people were employed. Another American firm, the Howe Machine Company, also had a works in Bridgeton from 1872. However, they must have found the competition from Singer too fierce, because in the late nineteenth century they began to manufacture bicycles and tricycles. There is an example of the latter in the National Museums' collection. In addition to the American firms there were a few small Scottish companies in the sewing machine business. Of these the most important was probably Kimball & Morton, of Glasgow, who made heavy machines for industrial use. The American factories used mass-production techniques unlike anything seen in Scotland before, but there seems to have been little attempt to learn from their example.

The manufacture of sugar machinery was an important branch of the west of Scotland engineering industry, but one which was not well-known, except to those directly involved. It began in the early years of the nine-teenth century with James Cook but others soon entered the field, although for some it was a short-lived interest. The genealogy of the industry is complicated. The most important firm was P & W McOnie, established in 1845 initially to supply spare parts for sugar machinery but soon moving on to the construction of new equipment for the sugar plantations. There was a split within the firm in 1848 and this gave rise to the two main strands in the industry's history. One led to the Mirrlees, Watson Company, which will be mentioned later in another context. The other became the Harvey Engineering Company after absorbing James Cook's old business. Other firms making sugar machinery included A & W Smith and Duncan Stewart & Company, both in Glasgow, and A F Craig & Company of Paisley. An indication of the scale of the industry is the fact that between 1851 and 1876 one firm alone, W & A McOnie (which later formed part of the Harvey Engineering Company), made 820 steam engines, 117 water wheels, 1,200 boilers, 1,650 cane mills and 169 evaporating pans for the sugar industry. While the West Indian market had been the most significant in the early days of the industry, other areas were growing in importance and many of the above items were for Java, Mauritius and Brazil. These figures also reflect the fact that at the time the average size of sugar mills was small, each serving a single sugar estate. In later years it was found to be more economical to build fewer but larger sugar factories serving whole districts, with the cane being brought by narrow-gauge railway systems. At the same time efficiency was improved by developments in the machinery which increased the amount of sugar extracted from the cane. With the equipment made by James Cook in the early years of the nineteenth century, the extraction rate was probably only about fifty per cent. One hundred years later something like ninety-five per cent of the sugar could be recovered.

Unlike sugar machinery the products of the structural engineers could hardly be more prominent. There were several important structural engineering firms but the best known was, of course, Sir William Arrol & Company of Glasgow, whose origins go back to 1868. They built the second Tay Bridge, the Forth Bridge and London's Tower Bridge, which were completed in 1887, 1890 and 1894 respectively. For a time all three contracts were running simultaneously. Arrol himself spent two days each week on the Forth Bridge and Tower Bridge sites, one day at the Tay Bridge and a day in the works in Glasgow. This was a punishing schedule. In addition to these spectacular jobs, the firm also built many less well-known bridges including various types with movable spans for navigation purposes. An example of these is the Temple Bridge over the Forth & Clyde Canal near Anniesland, Glasgow, a rolling lift bascule bridge built in 1931. In addition to bridge work, Arrol's made steel-framed buildings for engineering works and shipyards, and also large cranes.

There were several firms specialising in self-propelled steam cranes running on rails. The history of this type of crane, as it happens, offers a good example of a process which happened not infrequently in the engineering industry. The spread of expertise in a particular field was often the result of the movements of key people in the industry and if they did not move far the result was a cluster of firms in the same locality. Alexander Chaplin & Company was set up in Glasgow in 1849. Initially the range of products was fairly wide, but by the 1860s the firm had

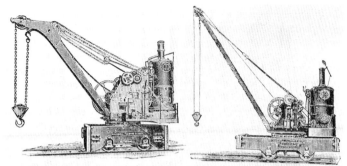

LOCO. and PORTABLE STEAM CRANE
For Contractors, Steel Works, Wharves, etc.

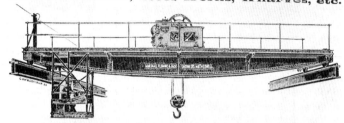

OVERHEAD TRAVELLERS,
Hand, Rope, Steam or Electric, up to 100 tons.

MARSHALL, FLEMING & JACK,
Motherwell, Scotland.

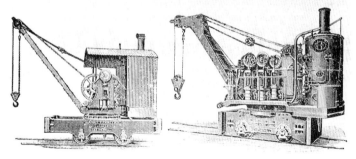

Marshall Fleming & Jack's advertisement in the catalogue of the 1901 Glasgow Exhibition. This Motherwell firm specialised in rail-mounted travelling cranes, and especially heavy duty cranes for steelworks. As shown, overhead cranes were also made.

acquired a special reputation for cranes and rail-mounted steam cranes in particular. A crane of this type was used to install machinery at the second Crystal Palace Exhibition in 1862. Working with Chaplin from 1858 to 1863 was George Russell. Confident that he could make better cranes than Chaplin, in 1865 he set up his own firm in Motherwell, George Russell & Company. About 1879 John Grieve left Russell to establish his own business, also in Motherwell. Ten years later there was a further breakaway from Russell's to create the firm of Marshall, Fleming & Jack, again in Motherwell. Yet another split occurred in 1907, when Alexander Jack went his own way and started up a new business, still in Motherwell.

The development of crane building in Motherwell was, in part, a consequence of the special needs of the local steel industry. Wherever there were important local manufacturing industries there were specialist engineering firms supplying their needs. Papermaking and printing were predominantly industries of the east of Scotland, although they were also carried on elsewhere, and the main engineering firms which served them were based in Edinburgh. There were two separate firms bearing the name Bertram, both making machinery for the paper industry. The brothers George and William Bertram started one firm in 1821, and the other was established in 1845 by another brother James Bertram. With

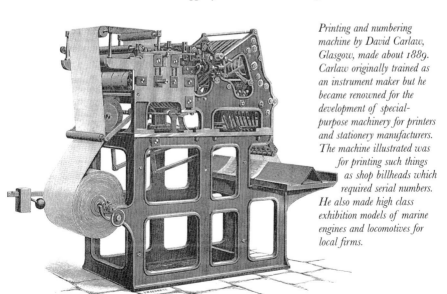

Printing and numbering machine by David Carlaw, Glasgow, made about 1889. Carlaw originally trained as an instrument maker but he became renowned for the development of special-purpose machinery for printers and stationery manufacturers. The machine illustrated was for printing such things as shop billheads which required serial numbers. He also made high class exhibition models of marine engines and locomotives for local firms.

Seven-colour linoleum-printing machine, 1889. This was developed by J Wright of the Kirkcaldy linoleum manufacturer Barry, Ostlere & Company, and built by J Copeland & Company of Glasgow. The linoleum was mounted on the large drum and the printing rollers, one for each colour, were carried on the frame to the left.

both firms engaged in essentially the same trade it is perhaps surprising that it was 1967 before they merged. Edinburgh was a major centre of the printing industry, but while there were makers of printing machinery in the city they were never able to meet all the needs of the printers and much machinery came from outside Scotland. Among the local firms were D & J Greig and Thomas Long. The Glasgow firm of David Carlaw developed numerical printing machines, which printed things such as tickets for transport undertakings who required progressive numbering of tickets.

In the north-east of Scotland there were makers of equipment for the Aberdeen granite industry. John M Henderson & Sons, founded in the mid 1860s, built aerial ropeways (sometimes referred to as 'Blondins'), steam cranes and steam-driven rock drills for the quarries. In addition they produced special lathes, saws and polishing machinery for working the granite into the required shapes. Another firm, George Cassie & Sons, also made stone-cutting and polishing equipment. The idea of using compressed air to operate rock drills and other tools came from the United States in the 1890s. Initially the tools were imported, but in 1905 the Consolidated Pneumatic Tool Company was set up in Fraserburgh. At the end of World War I the Bon-Accord Pneumatic Tool Company, making similar types of equipment, was established in Aberdeen itself.

A significant part of the engineering industry was devoted to the production of things needed within the industry. These ranged from 'semi-finished' items such as castings and forgings, to machine tools ('machines to make machines') and right down to the most basic items such as boiler tubes, bolts and nuts, and rivets. Many engineering companies had their own foundry and the name of their works often included the word 'foundry'. For example, the Kirkcaldy works of Douglas & Grant was known as the Dunnikier Foundry, although many activities besides the production of castings went on there. Likewise a blacksmith's shop in which wrought iron, and later steel, components could be forged to shape

Steam-driven rivet and bolt-making machine as made in the late 1880s by James Millar & Company, Glasgow. This firm was one of a number of specialist manufacturers of rivets, bolts, woodscrews and similar goods in Glasgow. They also manufactured and sold the machines used to produce them.

The foundry at the Kirkcaldy works of Douglas & Grant, early in the twentieth century. Moulds are being prepared in the sand using wooden patterns. When all the moulds are ready they will be filled with molten iron to produce the castings for machine parts.

was a common feature in engineering works. There were, too, jobbing foundries and forges to which work could be sub-contracted. However, the production of the castings and forgings needed for such things as the largest marine engines could be undertaken only by a handful of firms. A few of the larger engineering firms could produce these components for their own use and sometimes supplied other firms. In addition specialised firms grew up to deal with this class of work. Among them was the Govan Foundry Company which was established in 1867 specifically to produce heavy castings for the marine engineering trade. For the production of heavy forgings the Lancefield Forge, Glasgow, established in 1827, was for a time in a class by itself. Only the Lancefield Forge was able to produce the forgings for the stern frame (25 tons or 25.4 tonnes), paddle-engine crankshaft (31 tons or 31.5 tonnes) and propeller shaft (35 tons or 35.6 tonnes) for Brunel's *Great Eastern*. The other renowned forge in Glasgow was Parkhead. This was started in 1837 by Reoch Brothers. When they failed in 1841, first David Napier then Robert Napier became involved because they wanted a source of forgings for their marine engineering activities. However, Parkhead will forever be associated with the name of Beardmore. This became linked with the Forge in 1861 when the first William Beardmore joined William Rigby, who had acquired it in the previous year from his father-in-law, Robert Napier. Under William Beardmore junior, Parkhead Forge was to become the centre of a huge steel-making and engineering empire in the late nineteenth and twentieth centuries.

Large forgings, such as those turned out at Lancefield and Parkhead, could be produced only by the use of the steam hammer. The earliest type of hammer in widespread use was that patented by the Scottish-born engineer James Nasmyth in 1842, although it seems that he had sketched out a design in 1839. In the years which followed there were many attempts

Brunel's huge ship Great Eastern *of 1858 was built on the River Thames. It had two engines, one driving a propeller and one driving paddles. This shows the paddle-engine crankshaft, weighing 31 tons (31.5 tonnes). It was made by the Lancefield Forge, Glasgow, probably the only works in the country capable of producing a shaft of this size.*

to improve the steam hammer. The best which emerged from all this activity was that patented in 1854 by William Rigby who had been manager of Parkhead Forge since 1845 and had ample opportunity to study the vices and virtues of the various hammers. As noted above, Rigby later became owner of Parkhead and then the partner of William Beardmore. Manufacture of the Rigby hammer was taken up by Glen & Ross of Glasgow who made them in large numbers, and in a wide range of sizes. The firm subsequently became R G Ross & Son, a name which will appear again later.

In the early years of the nineteenth century the only machine tool in many engineering works was a lathe, sometimes made by the user. There was no way of producing a flat surface except by hand chipping with a hammer and chisel and then filing. At least one firm in this period was recruiting stonemasons for this task. When David Elder went to work for Robert Napier in 1821 the only machine tools there were two primitive

Rolling steel at William Beardmore's Parkhead forge, Glasgow, in 1966. This shows the cogging mill, in which the first stage of the rolling of the white-hot ingot to the required size was carried out. The mill was built by Duncan Stewart & Company, Glasgow.

The lathe is the basic machine tool, used to make round parts of every description from wooden chair legs to large, complicated machinery parts such as rotors for steam turbines. This huge lathe, mostly hidden by the employees of its maker, Thomas Shanks & Company of Johnstone, is clearly intended for the latter role! It dates from the early part of the twentieth century. (Dan Mackay Collection)

lathes with wooden beds and crude horizontal and vertical boring machines. Elder himself designed various new tools and these were made in the works. As the engineering industry developed, the general engineering firms began to include machine tools among their products. By the middle of the century there were a number of specialised machine-tool builders, and within this trade there was some further specialisation on particular types of tool. Notable firms were Thomas Shanks & Company of Johnstone, who were established in 1824 to make spinning machinery, but when this began to stagnate they successfully made the transition to the manufacture of machine tools for the growing engineering industry. They became builders of the heaviest lathes, drilling machines, slotting machines and cylinder boring machines to be made in Scotland. Another Johnstone firm, Loudon Brothers, made planing machines up to the largest size, although other types of machine tools were also made. John Lang & Sons, yet another Johnstone-based firm, was established in 1874 and for the first five years they built various different machine tools. Then they took the decision to build only one kind of tool, the lathe. These were made in a variety of types and sizes, but by developing their expertise and manufacturing capacity to suit this one type of machine they had the edge in a very competitive market. They were, too, responsible for one major technical innovation which was in time taken up by other machine tool makers. This was the use of gears with machine-cut teeth, which were more accurate and more durable than the cast gears used previously. Lang's claimed to be the first makers in Britain to use only machine-cut gears in their tools.

In the early days of boiler-making the plates and other iron parts were cut to size, formed to the required shape, and had the necessary rivet holes punched, all by hand. With the development of iron shipbuilding the quantity of material to be processed was greatly increased and there was a need for special machinery. Guillotines were built to cut plates and rolled sections (for frames and deck beams) to size. Bending machines shaped them and punching machines made the holes for the rivets which held the parts together. It was not uncommon for tools of this type to be driven by individual steam engines because they were scattered throughout the shipyard, unlike machines in a factory which were usually located close together and could therefore be powered by a system of belts and pulleys driven by a single engine. Similar machines were also used in the construction of bridges, tanks and gas-holders. There were several important makers of these tools, notably Craig & Donald of Johnstone, founded 1850 and one of the pioneers in this field. Hugh Smith & Company, founded in Glasgow in 1875, is significant as the only Scottish-based survivor and more will be said about the firm later.

Moving from the complications of machine tools to the simplicity of tubes, the first specialised manufacturer in this field was Andrew Liddle who began making gas, water and steam tubes in Glasgow in 1836. Other soon followed and by 1900 there were eleven west of Scotland firms in the trade. Among them was A J Stewart & Menzies, founded in Glasgow in 1860, which went on to absorb several of its rivals, and in the 1930s, as

HUGH SMITH & CO.,
Possil Engine Works, GLASGOW.

MACHINE TOOLS

Boilermakers and Shipbuilders.

BENDING, PUNCHING, SHEARING, PLANING, RIVETING, and DRILLING MACHINES, HYDRAULIC CRANES, PRESSES, Etc.   PUMPS and ACCUMULATORS.

. . . MAKERS OF . . .

Lambie's Patent Hydraulic Angle and Plate Joggling Machine.

HYDRAULIC MANHOLE PUNCH.

PATENT PLATE BENDING ROLLERS.

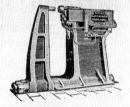

PATENT HYDRAULIC RIVETING MACHINE.

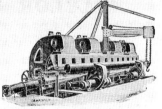

PATENT HYD. PLATE FLANGING MACHINE.

VERTICAL PLATE BENDING ROLLERS.

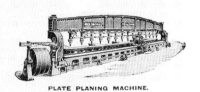

PLATE PLANING MACHINE.

This advertisement is from the catalogue of the 1901 Glasgow Exhibition. It illustrates some of the special plate-working machine tools used by shipbuilders and boilermakers. Hugh Smith & Company, established in the 1870s, was one of a small number of Scottish firms who made tools of this type and is the only survivor.

Town gas was made by heating coal in ovens known as 'retorts', about 10 ft (3 m) long. In a small gasworks the work of filling and emptying them was done by hand. Large works in the cities had hundreds of retorts and there was a strong incentive to mechanise the process. Dunbar and Nicholson's steam-powered machine was first developed about 1866 and the illustration shows an improved machine built a few years later by Henry Balfour & Company, Leven.

Stewarts & Lloyds Ltd, built a new steelworks and tube works at Corby in Northamptonshire. Other very basic items used in large quantities by the engineering and shipbuilding industries were rivets, and nuts and bolts. Their manufacture was sometimes regarded as useful training for young apprentices, but the fact that they were cheap labour was probably more important in the eyes of most employers. However, these essential fastenings were also mass-produced using special machinery by firms who made nothing else. For example, William Crosher & Company, Kinning Park, Glasgow, founded in 1840, was by 1888 employing seventy people. By 1900 there were at least fifteen firms in the west of Scotland engaged in this kind of work.

In considering 'who made what' in relation to the various firms and their products, it should be borne in mind that there was a considerable amount of sub-contracting and selling through agents, whose name might appear on the item rather than that of the maker. A note in the records of the Glasgow firm of A & W Smith, concerning an order, about 1865, for sugar plant (vacuum pan condensers and receivers) from Alberto Robinson, London, says, 'No tickets or name plates other than those of A. Robinson, London, to be put on the Condenser or Receiver nor any part thereof'. There were also firms, particularly in the machine tool trade, who manufactured some types of machines and bought in others in order to offer customers a wider range. The name prominently shown on the machine was not necessarily the actual maker!

What of the men who built the machines, the working engineers? Few of them left any record of their careers, but the following accounts from the writings of Alan Dunbar and Jimmie Houston give something of the flavour of the industry. The late Alan Dunbar, in his book *Fifty Years with Scottish Steam*, writes about his apprenticeship as a locomotive fitter at the Caledonian Railway's St Rollox works in Glasgow and his subsequent railway career. At that time St Rollox built new locomotives as well as repairing old ones and also dealt with carriages and wagons. By the time Alan Dunbar, aged fifteen, went to St Rollox works in the middle of World War I, the engineering industry had had long enough to become set in its ways and his experience would have been similar to that of apprentices elsewhere and at other times. The first morning, starting at 6 o'clock was something of a shock. The rows of dismantled locomotives, the general clutter and the gas lighting, made for a dismal scene. Like all apprentices Alan Dunbar began with a spell in the stores. From there he went to the section of the erecting shop where engine tenders were repaired and learned the use of the basic tools of the fitter's trade, the hammer, chisels and files by means of which parts were made to fit. As a new apprentice he inevitably had his leg pulled, being sent for 'the long stand' or some other such fool's errand! After working on tenders he went to a section repairing locomotives. There were six journeymen there and the charge hand made a point of attaching apprentices to each man in turn, rather than letting him become a general handyman for the squad. The men tended to specialise in different parts of the work on a locomotive and the apprentices therefore gained wide experience. Outside the main works there was a 'steaming shed' which carried out trials of new and repaired locomotives, and Alan Dunbar was sent there as a third year apprentice or 'improver'. He notes in passing that by this time his weekly pay had risen from 17/6d (87½p) as a first year apprentice to 57/6d (£2.87½p), including war bonus. His experience in the steaming shed made him want to work with operational locomotives rather than the dead ones in the works. He therefore moved from St Rollox to the running shed (in more recent terminology, the motive power depot) at nearby Balornock to complete his apprenticeship and it was in running sheds that he was to spend the rest of his career. He makes some interesting comments about his fellow apprentices. Some just wasted their time, learned little and were unlikely to be kept on when their time was out. There was one in particular whose main talent lay in knowing the unofficial ways of getting out of the works so that he could buy cigarettes, visit the bookie and undertake other such jobs for the men during working hours. But many were keen to learn and proud to be 'Caley men'. They were anxious to widen their experience too and when their time was out small groups would set off for the big wide world to gain experience elsewhere, working in perhaps as many as a dozen places in eighteen months before coming back to Springburn and settling down at one of the works but not necessarily the one at which they had served their time. There was no formal training organised by the company and, it seems, little encouragement to go and seek it elsewhere. However, apprentices like Alan Dunbar who wanted to learn and were not slow to ask questions, received much informal help and encouragement from the charge hands and older men.

In a different vein, the tale which follows comes from a long-standing and very observant friend of the author, the late Jimmie Houston, whose writings in the newsletters and magazines of the Scottish Traction Engine Society over many years have entertained and fascinated its members. This is a cameo portrait of one department in a single firm, but it illustrates well a number of interesting facets of life in the Scottish engineering industry. It might be described as a worm's-eye view. Although dating from the years of World War II, most of it could have happened at any time in the half century before and for a decade or two afterwards. Jimmie Houston served his apprenticeship with A & W Smith of Glasgow, who were well-known for their sugar machinery, but in 1932 they had acquired the firm of R G Ross & Son, makers of the Rigby steam hammer, and transferred production of this to Smith's own factory. Jimmie happened to be working in the hammer shop in 1943-4. Steam hammers were by this time being displaced by electrically-powered compressed-air hammers and the receipt of an order from the Great Western Railway (irreverently known as God's Wonderful Railway) for a small steam hammer for its Swindon works was quite an event. The hammer was duly completed and sent off in a railway wagon. It was back three weeks later with the complaint that it was uncontrollable. The hammer was then set up in the works for a demonstration. Jimmie remarks that the smiddy (blacksmiths' shop) in the works had electrically-powered hammers and these were driven by girls, none of whom had ever had any dealings with a steamer. It is interesting that the hammer drivers were women, but it was wartime and women were then doing many things which they would not have done in earlier years. Note that the hammers were not merely operated, they were 'driven' by hammer drivers. However, there was still in the works an old steam-hammer driver, Sam MacCallum, and he was brought along to try this allegedly uncontrollable beast. This he did, cautiously at first, but no problems were apparent and it was pronounced to be a 'clinking wee hammer'. Word was sent south and in due course a delegation arrived 'in the form of three "high hats" from Swindon, one large and impressive in a bowler hat and two underlings in soft hats and belter raincoats, no doubt the head blacksmith and his two charge hands'. Sam MacCallum showed them just how gentle an animal the hammer was, but they were unconvinced. After discussion with the management it was agreed that another driver would be found to do a demonstration. In Jimmie Houston's words:

Now at that time there was a firm in East Campbell Street near Glasgow Cross called Mathieson whose joiner hand tools, planes, chisels etc, were legendary. The blades of these tools were forged below very small hammers powered by real steam, some of them as small as 56 pounds [or 25 kg; this refers to the falling weight of the hammer] and reputedly driven by the smartest hammer boys on Clydeside. I'd been in the works once and was utterly amazed at the skill of these children. The school leaving age was only 14 at that time. It was said that they could deliver several hundred blows per minute to keep the necessary red heat in the thin tools. One of these prodigies was requested from Mathiesons and soon he appeared in our works, by which means of transport I know not. He was a typical Glasgow waif of that harsh economic era, small, undernourished, almost in rags with a large bunnet stuck on the side of his head. 'Bowler hat' and his cronies looked scathingly at the boy, but nothing daunted he got a piece of hardwood below the hammer and got himself accustomed to it by beating the hardwood, not too vigorously, into a pulp.

The men from Swindon could not take this young hammer driver seriously and when a billet of white hot steel (known simply as a 'heat') was brought from the furnace they drew nearer the hammer. Smith's foreman blacksmith, Jimmie Britton, who had hold of the tongs in which the heat was gripped, shouted the simple command, 'hit it' and

> all hell broke loose. Poor Jimmie had to handle himself smartly to turn the tongs quickly enough to avoid a foul blow on the corner of the square billet with the possibility of the tongs being knocked out of his hand and maybe killing someone. The blows rained on the billet with the rapidity of a rivet gun …. Hard hat & Co. jumped about ten feet in the air and leaped back out of the way ….
>
> The wee hammer boy laughed at them and danced around shouting, 'I shewed them whit a guid driver could dae, didn't I?'

No doubt he was given a small reward and went on his way, back to East Campbell Street. The hammer was sent off to Swindon a second time and no more was heard. Was it ever used, one wonders? That young boys were regularly employed on this kind of work is a sobering thought. The noise level would have been deafening and handling hot metal under a steam hammer could be dangerous if the almost instinctive communication between the blacksmith manipulating the tongs and the hammer driver failed momentarily. Once the hammer driver had started to hit it, verbal communication was virtually impossible. And these were 'the good old days'! Yet in people who have worked in such conditions there can be sensed a satisfaction and pride in having done the job well in spite of the difficulties. Distance lends enchantment, but it is hard to imagine that there could ever be similar feelings about assembling electronic components.

4
Workshop of the World

The west of Scotland in the nineteenth century has, with justification, been called the 'Workshop of the British Empire', but Scottish engineering products were also exported to many areas outwith the Empire. Moreover, there were engineering firms in other parts of Scotland who were also major exporters. Sales to England were also very important, although whether these should be regarded as exports depends on the political viewpoint! Products exported included steam engines to power factories, textile machinery, railway locomotives and rolling stock, mining equipment, sugar machinery, rice mills, iron and steel bridges, and prefabricated iron buildings. Most of the ships used to transport these products were also built in Scotland and many were operated by Scottish-based shipping companies. Ocean-going ships built for foreign-owned shipping lines were an important element in the export trade, as were the numerous small vessels built for service on lakes and rivers throughout the world.

Among the industries most heavily dependent on exports were those supplying what might be described as 'plantation machinery', for crops which were grown only overseas and largely processed where they were grown. For the Scottish engineering industry sugar was the most important of these, but rice and coffee were also significant. The firms in these trades exported a high proportion of their output. Although some sugar machinery was supplied to refineries in Britain, the bulk of it was built for overseas markets. A list published in 1901 of the destinations of complete cane-sugar factories supplied in the previous few years by one Glasgow firm, Duncan Stewart & Company, shows the wide-ranging nature of the industry. Factories went to Brazil, Mexico, Egypt, India, Mozambique, India, Australia, British West Indies, Honduras, Japan and the Straits Settlement. Five factories went to Mexico and more than one to several of the others. In the early days of the industry the mills were based in individual sugar-cane plantations. While the mills were numerous and therefore relatively small, the sheer number in operation helped to steady the demand on the machinery builders for both new machinery and for spare parts. However, with the coming of the centralised factories serving a whole district the requirement was for fewer but much larger mills. In 1953-4 the Paisley firm of A F Craig received only orders for a series of relatively small jobs, including the supply of spares for carpet looms, some sub-contract work for other makers of sugar machinery and a few marine boilers for shipbuilders in Paisley and in Sweden. Then in May 1955 there came orders for two complete sugar factories, to be built in the Punjab. These were large, each factory costing £450,000 and having a cane processing capacity of 1,000 tons (1,016 tonnes) per day. To cope with such large fluctuations in workload the firms in the sugar machinery industry sub-contracted work when necessary.

Other types of machinery were made for use in plantations both within the Empire and in the wider world. Robert Douglas of Kirkcaldy (later Douglas & Grant) was one of the leading Scottish builders of steam engines for powering factories. The firm also developed a thriving business in the manufacture of rice-milling machinery for Far Eastern countries such as Burma, Siam (Thailand) and Indo-China (Vietnam). Robert Douglas first became involved in this trade as the result of a visit to India in the 1860s, in connection with steam engines for jute mills being built in Calcutta. It was suggested that it would be worthwhile to visit Burma where there was some interest in the possibility of building steam-powered rice mills. Douglas acted on this hint and began to make machinery for rice processing. Not only was he able to broaden the base of his business by entering a new field, but he had an additional market for his established products, factory steam engines. Although the manufacture of steam engines ceased about 1930, rice-milling plant was still being made many years later.

The Aberdeen firm of William McKinnon & Company was another maker of rice mills, but also built machinery for coffee, cocoa and sugar plantations. Set up in 1798 as a foundry, they began to make plantation machinery in the 1860s. More than a hundred years later a feature on the firm, in *The Scotsman* for 31 May 1967, indicated that some ninety per cent of the output was exported and the remainder was mainly items for the local granite and paper industries. The comment was made that the practice of ex-apprentices from the firm going abroad to maintain the machinery built in Aberdeen had recently died out: 'There is a trend for the emerging

ABOVE: *An early twentieth-century sugar mill by A & W Smith of Glasgow. The mill is shown erected in the makers works, complete with the two steam engines and gearing by which it was driven.*

LEFT: *Sir William M'Onie (circa 1813-1894) played a leading part in the development of the sugar machinery industry in Glasgow. He also found time to become involved in civic affairs, becoming a town councillor in 1867 and Lord Provost from 1883 to 1886. The Baillie (a sort of local nineteenth-century Private Eye) from which this portrait comes, had a poor opinion of his ability to control unruly councillors, although praising his talents in other directions!*

countries to appoint their own men to mind the machines, which, mercifully, they still import.' The movement of men from manufacturers to the plantations for a spell, to operate and maintain the machinery which they had built, was also a feature of the sugar engineering firms. Jimmie Houston, who served his apprenticeship with the Glasgow firm, A & W Smith, told the writer that the usual practice with this firm was for men to go on a three-year contract to erect the new machinery and run the mill for three seasons. Sometimes works engineers were also sent out to keep old mills going. There was one occasion on which men went out to South America, to a somewhat decrepit mill nearing the end of its working life. This was driven by a Corliss-valve steam engine, a type with which they were not familiar. Jimmie recalls running an informal correspondence course on the peculiarities of Corliss engines and the operating problems which might be encountered! Experience in the mills was valued, and men coming back to Glasgow were sure of a job at the works.

In the early years of the locomotive-building industry in Glasgow, the main customers were the numerous British railway companies. As Britain had led the world in the development of railway systems this was only to be expected. When the first railways were built in other countries British engineers and contractors were heavily involved and more often than not it was from Britain that the locomotives and other equipment came. From Neilson's original works in Finnieston sixty-eight out the total of 429 loco-motives built between 1843 and 1862 went overseas. The first export locomotive was for Cuba in 1852 and before the end of the decade there were orders from Canada and India. As the development of railways overseas gathered momentum exports grew in importance. The countries of the British Empire were major markets, but Glasgow-built locomotives were to be found in almost every country with a railway system. The African colonies, together with India, were specially important markets. One of the few countries in which British makers were never able to establish even a toe-hold was the United States, where a large home-based industry rapidly became established. An export order which was of particular interest, because of the circumstances in which it was placed, concerned the supply of five express locomotives to Belgian State Rail-ways in 1898-9. The Caledonian Railway had introduced a new design of express locomotive in 1896. The first of these was named *Dunalastair*, after the estate of the then deputy chairman of the company and this soon

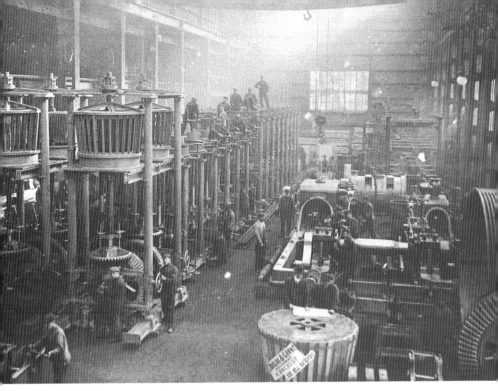

The erecting shop at the works of Douglas & Grant, Kirkcaldy, in the early years of the twentieth century. The machines on the left of the picture are rice mills. On the right is a Corliss-valve steam engine. This photograph is of particular interest as it includes people, who rarely feature in pictures of workshop interiors.

became the collective name for the class. The Dunalastairs proved to be one of the best late nineteenth-century locomotive designs and their exploits were well-publicised. The first of an improved version which came to be known as the Dunalastair IIs appeared late in 1897. Early in the following year the chief engineer of Belgian State Railways wrote to the Caledonian saying, in effect, that they would like some of these engines please. John F McIntosh, the Caledonian Railway's locomotive super-intendent, was authorised by the Board of Directors to supply drawings and arrange for the construction of the locomotives in Glasgow under his supervision. The builder chosen was Neilson, Reid & Company. This was the long-established Neilson & Company, renamed in 1898 to reflect, belatedly, the fact that control of the company had passed in 1876 from Walter Montgomerie Neilson to James Reid and his family. Despite the importance of the Caledonian Dunalastairs in Scottish railway history none has been preserved in Britain, but one of the Belgian locomotives is in a museum there.

The raw materials used by much of the British textile industry was imported, notably cotton and jute. Fearing the establishment of competing factories overseas the British government banned the export of the newly developed power-driven textile machinery between 1782 and 1843. Despite the ban, machinery was sent overseas. For example, in 1805 the Swiss firm of Escher Wyss & Company began the construction in Zurich of a cotton spinning mill with English machinery. After the ban was lifted cotton manufacture grew up rapidly overseas. There were mills in India within ten years and in Japan by the 1870s. Douglas & Grant of Kirkcaldy was among the suppliers of steam engines to drive these Indian cotton mills and some of the engines which they made for service in India were far more powerful than anything for the home market. One of 1,000 hp (764 kw) installed in a Bombay cotton mill in 1876 was said at the time to be the largest steam engine ever made in Fife. Around ten years later another large engine was built for the Gordon Spinning and Manufacturing Company, which from its name one might suppose to have been in Scotland. In fact it was another Bombay mill but perhaps one which was Scottish-owned. Similarly jute mills were set up in India, with Scottish money, management and machinery, much of it from Dundee. For these mills many engines came from Dundee and Kirkcaldy.

A lot of overseas organisations which bought machinery from Scottish engineering firms had strong financial links with Scotland or with Britain as a whole. A good customer of the Glasgow locomotive builders was the British-owned railway system in Argentina. The Tharsis Sulphur & Copper Company was set up in 1866 with Glasgow money, to mine the pyrites deposits in southern Spain. The chairman was Charles Tennent of St Rollox chemical works, where the pyrites was used as a source of sulphur. Glasgow-built locomotives were used to transport the minerals to the port for shipment. An early steam excavator was bought from Andrew Barclay of Kilmarnock to speed the extraction of the pyrites, but it seems that this pioneering machine was not entirely successful. As already mentioned, it was the fact that the West Indian sugar plantations were owned by Glasgow merchants which gave the local sugar-machinery makers a start. Many Clyde shipbuilders were financially involved in

During World War I the North British Locomotive Company built over 1,400 locomotives in addition to many different types of war material. They were supplied to British and French railway companies, who desperately needed them to cope with the greatly increased traffic as a result of the War, but many were also built for the company's regular overseas customers. The illustration shows one of ten for South African Railways 3 ft 6 inch (1,067 mm) gauge system.

shipping companies. Peter Denny, of the Dumbarton firm William Denny & Brothers, was an major shareholder in the Irrawaddy Flotilla Company which was formed in 1876 to operate river services in Burma and his firm built 250 vessels for the company. Direct sales to foreign governments were also significant, especially for warships. In 1864 three armoured frigates for the Turkish Navy were ordered from Robert Napier & Sons. On the face of it this was an order placed by a foreign government with a private contractor, but in reality it was almost certainly a matter between the British and Turkish governments. These ships represented the latest technology and the Royal Navy had received comparable vessels only a year or two earlier. Napier's could not have gone ahead with these vessels without the cooperation of the British government and this would not have been forthcoming unless the sale was compatible with British foreign policy in the region. In 1868 a very fine model of one of the ships, the *Osman Ghazi*, was presented by the shipbuilder to the Museum of Science & Art which later became the Royal Scottish Museum and is now part of the National Museums of Scotland.

Scottish engineering products built for export were often very different from those for the British market. For example, many of the locomotives built for service in South Africa were much larger and heavier than any which ever ran in Britain, in spite of the fact that they ran on track with a gauge of only 3 feet 6 inches (1,067 mm) instead of the standard of 4 feet

$8\frac{1}{2}$ inches (1,435 mm). The result was a dramatic difference not only in appearance but in the way they were designed and built.

While some significant examples of Scottish-built machinery have been preserved in Scotland or elsewhere in Britain, important items also still exist overseas. Glasgow-built steam locomotives are in use or preserved in museums. There are far more Scottish-built historic ships preserved abroad than there are in Britain. There is sugar machinery in regular use and almost certainly even more abandoned in former plantations, now reclaimed by the jungle. The vagaries of survival have meant that some important Scottish engineering firms are represented only by products remaining abroad. Taking the nineteenth-century beam engine as one example, the only known complete McNaught compound beam engines by the Canal Basin Foundry, Glasgow, maker of the first engine of this type, are both preserved in Dublin. To see an engine by the Kirkcaldy firm of John Key it is necessary to go to the other side of the world, to Auckland, New Zealand, where one built in 1877 is preserved in a former waterworks which is now a museum. However, the idea of 'restitution of cultural property' is beginning to extend beyond the world of art and ethnography. Several locomotives, including examples built in Glasgow by the North British Locomotive Company, are now back in Britain at the end of their working lives. In 1970 Brunel's iron ship *Great Britain* was brought from the Falkland Islands to Bristol where it was built, and the sailing ship *Glenlee*, built at Port Glasgow in 1896, came back to the Clyde in 1993. Perhaps one of these days a Glasgow-built sugar-cane mill will be repatriated for display in Glasgow!

5
New Ideas
for a New Century

After the water-powered phase of the Industrial Revolution in the late eighteenth and early nineteenth centuries, the pattern of development was largely shaped by dependence on the steam engine. Each factory was powered by its own steam engine or engines, raw materials and finished products were moved by steam-worked railways. With the dependence on individual power units and rail transport came a strong tendency towards the concentration of factories in particular locations. Conditions which suited one were suitable for others in the same trade and this contributed to the rapid expansion of towns and cities. The late nineteenth century saw the birth of a number of new technologies which were to result in profound industrial and social changes in the twentieth. None were more important than the development of the internal combustion engine and the creation of a nationwide electricity supply system. The internal combustion engine changed the pattern of transport, significantly reducing costs and bringing the flexibility of the motor lorry and private car. With the growth of the electricity supply industry almost all industrial power is now supplied via the national grid. The cost is lower in real terms and much more nearly uniform than it was when each factory had to produce its own power. The choice of location for a factory is therefore less constrained than it was in the nineteenth century. These were not the only examples of new technology at the turn of the century, but they were by far the most significant in their long-term effects. This chapter therefore looks at the response of Scottish industry to the opportunities created in these new fields.

The steam engine is an external combustion engine because the burning of the fuel takes place externally, in a separate boiler, rather than in the power-producing cylinder. The idea of an engine in which combustion takes place inside the cylinder is almost as old as the steam engine, but as often happens in engineering there was a big gap between the idea and the workable machine. It was 1860 before there was a marketable internal combustion engine, using town gas as fuel. However, a German inventor,

Nicolaus August Otto, is usually credited with the invention of the four-stroke operating cycle, on which most modern engines run. Otto's first engine appeared in 1876 and it quickly became apparent that a great breakthrough had been achieved. By 1889 some 40,000 engines had been built. The four-stroke engine has a power stroke only every second revolution. It occurred to more than one engineer that it ought to be possible to obtain more power from an engine which had a power stroke in every revolution, that is a two-stroke engine. Moreover, the fact that Otto had patented the four-stroke cycle provided an additional incentive to explore this avenue. One of the first satisfactory two-stroke gas engines was the work of Glasgow-born Dugald Clerk, later Sir Dugald. His first engine was built in 1879 and two years later it was in commercial production at the Glasgow works of Thomson, Sterne & Company. Clerk was one of the first British engineers to study the workings of the internal combustion engine in a scientific way. Like many other talented Scottish engineers, he found the road to England irresistible. He had joined Thomson, Sterne & Company in 1877 and moved in 1885 to the large Birmingham engineering firm of Tangye Brothers.

Although the first gas engines were small, larger engines were soon developed. For these, town gas was too costly and they used either a cheaper fuel made for power purposes in special gas producers, or waste gas from the furnaces of iron and steel works. The design of these large engines exhibited much ingenuity and one of the most interesting of them was the Oechelhauser which was developed in Germany in the closing years of the nineteenth century. This was a two-stroke opposed piston engine, that is one in which each cylinder had two pistons moving simultaneously in opposite directions. In 1903 William Beardmore & Company, Parkhead, Glasgow, acquired a manufacturing license for Britain and the Colonies. By this time Beardmore's had begun a period of rapid expansion and diversification from their core business as steelmakers and producers of heavy forgings and castings. They were in the process of adding a very wide range of engineering products and also becoming shipbuilders. The intention was that the gas engines would be built at the works of Duncan Stewart, London Road, Glasgow, by then a Beardmore subsidiary, and also at Beardmore's new shipyard and engineering works at Dalmuir to the west of Glasgow. The first engine was made at London Road to power a small rolling mill at Beardmore's

Gas engine made by Pollock, Whyte & Waddel, of Johnstone. This firm was one of the few Scottish makers of gas engines. The 4 hp (3 kw) engine illustrated was shown at the Brussels Exhibition of 1897.

Parkhead works. Unfortunately things went badly wrong with a contract awarded in 1904 to supply engines for a power station in Johannesburg. The engines could not be made to run reliably and there was a substantial loss on the contract. Few if any further engines of this type were built by Beardmore's or Stewart's. Yet the basic concept was sound and other people in other places made opposed-piston engines work very well.

The history of the motor industry in Scotland has recently been told by Alastair Dodds in *Making Cars*, another in the 'Scotland's Past in Action' series. It is a somewhat gloomy tale, relieved only by the success of Albion Motors. Founded in 1899, Albion grew rapidly and soon became specialist makers of commercial vehicles. With the outbreak of World War I Albion's production became a vital part of the war effort and by the end 5,594 lorries had been built for military use. The likely importance of motor transport in wartime was becoming recognised even before 1914, and at the same time there was growing realisation that the new-fangled flying machines might also have an important role. This fitted in with William Beardmore's expansive outlook at that time. Manufacturing licenses were obtained from continental firms for both aircraft and aero engines in 1913. The aircraft chosen was the German-designed DFW (Deutsche Flugzeuge Werke), but little was done with this. The engine license, for which Beardmore is said to have paid £10,000, was to prove much more important. Designed by a certain Dr Porsche (a name later to become well-known in the motor car world) of the Austro-Daimler works in Vienna-Neustadt, it was one of the best engines available to any of the belligerents in the early years of the war. The total production was over 3,000 engines, a considerable achievement. Most were built at the almost new Dumfries factory of Arrol-Johnston, the car manufacturing firm which had been part of the Beardmore empire since 1905.

The gas engine was the first internal combustion engine to gain widespread acceptance, but a gas supply was not available everywhere. There was therefore interest in the development of engines using oil as a fuel. Several successful oil engine designs were evolved from the mid-1880s onwards, and one or two Scottish firms, such as Allan Brothers of Aberdeen, became involved in their manufacture. These early oil engines required the use of a blow-lamp to pre-heat part of the cylinder, so that the oil was vapourised and would ignite. However, there can be little doubt that the most important internal combustion engine is the diesel engine. It is now the usual power unit for ships, non-electrified railways, heavy lorries and buses. In addition the diesel has many applications in industry. While most cars are still powered by petrol engines, diesel engines are increasingly being used.

Dr Rudolph Diesel was an academically trained engineer, born in Paris of German parents, and trained in Germany. The essential feature of the diesel engine is that air is compressed in the engine cylinder to such an extent that the temperature becomes high enough to ignite fuel oil as it is injected into the cylinder, without the need for any pre-heating or additional ignition device. Development was supported by the German firm Maschinenfabrik Augsburg, later known as MAN. The first engine was built in 1893. This was not a success but it provided useful information and experience. Further development engines were built and tested and by 1897 sufficient progress had been made to enable engines to be offered for sale. In addition, licenses to manufacture the engine were made available. One of the early licensees, and the first in Britain, was a Glasgow firm, the Mirrlees, Watson & Yaryan Company (from 1900 the Mirrlees, Watson Company). They had a 20 hp (15 kw) engine running in 1897, but there were problems with the fuel injection system which they were not able to cure at the time. Following the appointment in 1901 of Charles Day, who had previously worked for some of the major English builders of both steam and gas engines, development work was restarted.

The original engine was modified and two new engines of 35 hp (26 kw) were built. After initial trials they were used to drive machinery in the firm's works. It soon became abundantly clear that here was a reliable and extremely efficient engine, an invention of very great importance. Had

The first British-built diesel engine. In 1897 Mirrlees, Watson of Glasgow obtained a license to build diesel engines and had their first engine running in the same year. There were problems and work was stopped for a time, but by 1902 a modified version of the engine was running satisfactorily and further engines were being built.

the firm continued manufacture in Scotland, a major new branch of the engineering industry would have been created. However, it was not to be. Instead the firm of Mirrlees, Bickerton & Day was set up in 1907 to manufacture engines on a large scale in a new factory at Stockport, south of Manchester. In 1890 Bickerton had established the National Gas Engine Company of Ashton-under-Lyne, a few miles east of Manchester. Day came originally from the Stockport area and had worked there for a time with a local builder of gas engines, J E H Andrew & Company. With these local connections, plus a workforce in the area experienced in building internal combustion engines, it is not surprising that Stockport was chosen for the new works. It is still building diesel engines as Mirrlees Blackstone Ltd, part of GEC-Alsthom.

William Beardmore & Company made a serious effort to develop various types of diesel engine, for aircraft, ships, stationary use and road vehicles, over a lengthy period. As one of the largest Scottish industrial firms they should have been able to make a success of this venture. The company had serious financial problems, however, and could not devote the necessary resources to the work. Among the engines built was the *Tornado*, a 585 hp (435 kw) eight-cylinder unit. This was the type installed in the airship R101 which crashed in France with heavy loss of life in 1929. Beardmore's diesel-engine department was finally closed in 1937. It had been a brave pioneering effort to produce a range of modern, quick-running and relatively lightweight engines. With the benefit of hindsight the product range was too large. Under-developed and unreliable engines were therefore being put on the market. The situation was not helped by the fact that the firm had several other new ideas under development at the same time. A few of the smaller Scottish engineering firms had some success with diesel engines for land use, including Alexander Shanks & Sons of Arbroath and C F Wilson & Company of Aberdeen, but in the longer term they could not compete with the major English producers.

The most important application of the diesel engine is now ship propulsion. Just as Stephenson's *Rocket* brought the steam locomotive to the world's notice, although it was by no means the first, so the success of the Danish-built ship *Selandia* focused the marine world's attention on the diesel engine. Completed in 1912, with twin engines totalling about 2,500 hp (1,850 kw), this was the first large ocean-going diesel-engined vessel. It was built and engined in Copenhagen by Burmeister & Wain for the Copenhagen to Bangkok route operated by a Danish firm, the East Asiatic Company. A sister ship, the *Jutlandia*, was completed a few weeks later by Barclay, Curle & Company on the Clyde. They also built the engines, which were of Burmeister & Wain design. The large marine diesel was rapidly accepted on the continent, but British ship owners were slow to take it up. There were only two attempts to produce 'home-grown' Scottish engines. In 1914 the North British Diesel Engine Works Ltd was set up in Glasgow specifically to design and build large marine diesel engines. The firm was taken over around 1926 by Barclay, Curle & Company, who may well have had a financial interest in it from the beginning. Scott's Shipbuilding & Engineering Company of Greenock also tried to enter the field after World War I. Engines were installed in a small number of ships in the 1920s and early 1930s, but the products of neither firm proved to be sufficiently reliable to encourage wider use.

British firms were not able to devote very much attention to new ideas, such as the diesel engine, during World War I. But continental engine builders, such as Burmeister & Wain of Copenhagen and Sulzer Brothers of Winterthur, Switzerland, who had taken up the diesel engine enthusiastically at an early stage, continued active development work. As a result they were able to build up a technical lead which they still retain. By the time the marine diesel was generally accepted in Britain as the engine of the future, the Scottish firms had little choice but to build the designs of others, under license. These included the only successful British engine, the opposed-piston design developed by William Doxford & Sons of Sunderland, as well as Continental designs. Although there was no successful Scottish design of a large engine, experience with smaller engines was more satisfactory. The Bergius Car and Engine Company, founded in Glasgow about 1904, built a car which they called the 'Kelvin'. Only a few, perhaps about fifteen, were built before production stopped in 1907 and attention was switched to small marine engines. Another successful Glasgow-built engine was that made by Gleniffer Motors Ltd, set up in 1913 and later renamed Gleniffer Engines Ltd. Both firms built petrol, paraffin and diesel engines suitable for smaller vessels. Despite the recession between the wars, there was a market for engines suitable for fishing boats and other small craft. Many of the older fishing boats, built as sailing vessels, were having engines installed and the Scottish designs were good enough to win a significant share of the business.

The 1930s also saw the diesel engine become a serious alternative to the petrol engine in heavy lorries and buses. As early as 1909 Rudolph Diesel himself had carried out the first trials in a road vehicle but the engine was too heavy in relation to its power. During the 1920s, when diesel engines were becoming lighter and more powerful, some Continental vehicle makers were offering diesel-powered buses and lorries. By 1930 there were several trial installations operating in Britain and within a few years it had become the normal power unit for heavy vehicles. Unfortunately Scotland's only sizeable vehicle builder, Albion Motors, had seriously misjudged the potential of the diesel and had not attempted to produce their own engine. When customers began to specify diesel power they had to fit engines bought in from outside firms, such as Beardmore's unit of doubtful reliability and, with much happier results, the Manchester-built Gardner engine.

As in the case of the internal combustion engine, German engineers played a major part in the practical application of electricity. The United States was also an early player in this field. Before long the major German and American companies had established subsidiary companies in England. The start of electricity supply for general use in Britain dates from the early 1880s. By 1889 there was a supply in Glasgow. This was operated by the local firm Muir, Mavor & Coulson, a pioneering company which had begun generating electricity some years earlier, providing a dedicated supply to the General Post Office. In 1892, by which time the firm had thirty-seven customers, the operation was taken over by Glasgow Corporation. This was an industry on the verge of explosive growth. In the first year under Corporation ownership current was supplied to 108 users. In 1903 the number of consumers had increased to over 7,000 and ten years later there were almost 28,000 customers. This was for a product which, by present day standards, was very expensive indeed.

British manufacturers of electrical machinery faced serious competition from overseas firms and their British subsidiaries. The strength of this competition is well-illustrated by the equipment chosen for Pinkston power station built in Glasgow at the turn of the century. This was built specifically for the tramway system, rather than for general public supply, and was one of the largest generating stations of its time. The consulting engineer was an American, H F Parshall. Virtually all the new technology,

Pinkston power station, Glasgow, which supplied current to the municipal tramway system from 1901. The important pieces of equipment, the main generators and some of the steam engines which drove them, came from the United States. The picture shows the plant which condensed the exhaust steam from the engines. This was made by the Mirrlees, Watson Company of Glasgow, with the electric motors to drive the various pumps coming from the Edinburgh firm of Bruce Peebles.

the electrical equipment, came from the United States. The orders for the four main and two auxiliary generators were placed with the British Thomson-Houston Company of Rugby. This was an offshoot of an American firm, the General Electric Company, and it was they who actually built the machines. Two of the four 4,000 hp (2,500 kw) steam engines which drove the main generators were also American, while the other two were made in Bolton, Lancashire. The Scottish contribution was confined to a pair of 800 hp (600 kw) auxiliary engines from Duncan

Stewart of Glasgow, plus low-technology items such as the condensing plant, coal bunkers and coal-handling equipment. It must have been well-understood that local contractors were simply not up to the major electrical parts of the job, for the only thing which caused any controversy, within or outside the Corporation, was the placing of the order for steam engines with the E P Allis Company of Milwaukee. Although the Pinkston engines were big as land engines went, they were less than one third of the power of the largest marine engines being built on the Clyde and the idea that Glasgow, of all places, should have to buy steam engines from abroad was clearly anathema to many people.

In order to survive in the face of such competition from the multinational companies British firms tended to specialise in areas where they were not subject to the full force of competition from the giants of the industry. Mavor & Coulson (formerly Muir, Mavor & Coulson), in addition to organising an electricity supply in Glasgow, made generators and electric motors during the 1890s. At the beginning of the new century they had sufficient confidence in the future to build and equip a new factory which was, in the opinion of one contemporary commentator,

> the most important of its kind in Scotland, and in facilities for the rapid, economical, and accurate production of high-class electrical machinery is not surpassed by the best and most recent factories in the country.

While the market for large machines for central power stations was increasingly dominated by a few firms, there remained a substantial and indeed growing demand for smaller machines for general industrial use. Mavor & Coulson built machines for this market. From 1897 they began to specialise in equipment for the mining industry and their first electrically-driven coal cutter was appropriately named the 'Pick-Quick'. This was a well-judged move and the firm became world leaders in the field.

An interesting Scottish electrical engineering firm is that established by David Bruce Peebles. He was born in Dundee and served his apprenticeship with a local firm, Umpherston & Kerr. Thereafter he worked for a spell in England, before returning to Scotland as a partner in an Edinburgh gas-meter maker, Fullerton & Company. About 1866 he set up his own business in the city to manufacture gas meters, burners and governors. These soon became very highly regarded within the gas

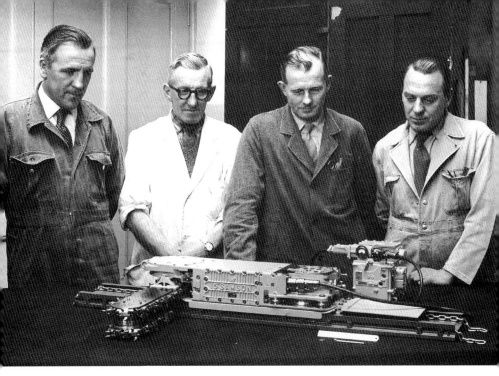

The Royal Scottish Museum, almost since its beginnings as the Industrial Museum of Scotland in 1854, had a workshop in which models were made of machines illustrating contemporary engineering practice. This model, one quarter full size, of an electrically-powered coal cutter by the Glasgow firm of Mavor & Coulson was finished in 1963. It was the last such model to be completed in the Museum. The members of the workshop staff involved in building the model are, from left to right, John Graham, Stuart Blackley, Bob Blackley and Jim Mason.
(A G Ingram)

industry. It is not every engineer or businessman, well-established and indeed famous in a major field, who is able to discern the significance of important developments outwith their industry. Bruce Peebles was one who had the necessary vision and in 1897 he set up an electrical department. David Bruce Peebles died in 1899, but he had made the right decision for the business and the electrical department grew rapidly. To help develop this side of the firm, a license was obtained in 1903 from the Hungarian firm of Ganz & Company, Budapest, for the use of their patents relating to alternating current machines. In time the gas and electrical businesses were separated.

The British Electric Plant Company, set up in 1904 at Alloa, was another major electrical manufacturer. In 1919 there was an amalgamation with the Harland Engineering Company of Manchester. Although the name of the latter was used for the combined firm, the Alloa works soon became the main manufacturing site. Products included pumps as well as electrical machinery. In the years after World War II the manufacture of water turbines was started, so that complete turbine and generator units could be built for the new power stations being developed by the North of Scotland Hydro-Electric Board. The firm is now part of Weir Pumps Ltd.

An early development in the electrical field was that of shipboard installations. The benefits of electric lighting instead of candles and oil lamps need no emphasis. Several firms in the Glasgow area took up the manufacture and installation of electrical systems on board ship, among them Mechan & Sons of Scotstoun, Glasgow. The firm began making electrical items in 1886 and within a few years they were undertaking complete installations of lighting and ventilation systems, telegraphs and telephones, searchlights and other equipment.

The Glasgow instrument maker James White, and his association with William Thomson (later Lord Kelvin) was important. Within a few years of starting up as an instrument-maker in 1850, White was making electrical apparatus for experimental and demonstration use. The firm became the only significant Scottish maker of electrical instrumentation, with a wide range of products, invented by Kelvin, for use in laboratories and in power stations.

In looking at the histories of these two key industries of the twentieth century the conclusion seems inescapable that, although there were important successes in Scotland, there were also too many missed opportunities. Fully developed, they could have gone a considerable way to offset the economic effects of the decline of the old industries and might indeed have made that decline less catastrophic.

6
The Wind of Change

The inadequate response to new technology has been outlined in the previous chapter. This was a British problem, rather than a particularly Scottish one. In 1880s and 1890s Scottish engineering and shipbuilding might, on the face of it, appear to be have been unchallenged world leaders still. The Glasgow locomotive builders exported seventy per cent of their output. Almost all the world's output of machinery for crushing and refining cane sugar came from Glasgow. British, and particularly Clyde shipbuilders and marine engineers, were the best. The domination of world markets by British shipbuilders is indicated by a list published in 1884, showing all large ships (over 4,000 gross tons) which had been built worldwide up until then. There were 138 in total, and of these only ten had been built outside Britain, seven in France and three in the United States. Of the British-built ships, over half came from the Clyde. The world's shipping companies came to Britain, and very often to the Clyde, for their largest and most important ships. However things were changing. Shipbuilding and marine engineering industries were being developed in overseas countries. Among them was Germany, whose major shipping lines, North German Lloyd and the Hamburg American Company, had long been customers of British yards. By 1890 German shipyards were able to meet all their needs. Germany was by no means the only country becoming independent of British technology. William Denny & Brothers of Dumbarton built a number of state-of-the-art, high-speed ships for the Belgian government's Ostend-Dover service, between 1888 and 1892. In the next few years further ships of similar design were added to the fleet, but these were built in Belgium. This increasing self-sufficiency in what had been important export markets should have caused concern among British shipbuilders, but there was little evidence of this. Privately, however, they were perhaps more concerned.

A wider awareness of what was happening did not come until 1897, when the Vulcan shipyard at Stettin produced for North German Lloyd the *Kaiser Wilhelm der Grosse*. This transatlantic liner was larger, more powerful and faster than anything else in the world, including the Cunard Blue Riband holders, *Campania* and *Lucania*, which had been built by the Fairfield yard at Govan in 1893. Further high-speed transatlantic liners for both North German Lloyd and the Hamburg American line were built by Vulcan in the next few years.

Questions were asked. Did Britannia still rule the waves? Something had to be done! In due course and with massive government assistance for both construction and operating costs, the turbine-powered *Mauretania* and *Lusitania* were completed in 1907 for Cunard. The former was built by Swan Hunter on the Tyne and the latter by John Brown, Clydebank. Although the German ships conclusively demonstrated that Britain's long-established technical superiority should never again be taken for granted, it is interesting to note that they incorporated the products of several British firms. Among them were the steering gear by Brown Brothers of Edinburgh, pumps from G & J Weir, Glasgow, and anchor windlasses and winches from Napier Brothers of Glasgow.

Not only in shipbuilding was it becoming apparent that conditions were changing. The acquisition by the Belgian State Railways in 1898-9 of five express locomotives, to the Caledonian Railway's Dunalastair II design, was mentioned in Chapter 4. In later years over 200 further engines to the same basic design were acquired, but all were built in Belgium. By the turn of the century the British locomotive-building firms, who had little experience of competition in overseas markets except amongst themselves, were faced with a serious threat from American firms and from Baldwin & Company of Philadelphia in particular. Germany was also becoming a major competitor. In 1900 Baldwin's had built over 1,200 locomotives, far above the combined capacity of the three Glasgow firms, and their output was still increasing. Expansion of the American railway system was slowing down and Baldwin's were looking for other outlets for part of their huge production capacity. They tried to break into the traditional markets of the British builders and had some success, particularly in India. Baldwin's even managed to sell a few locomotives to three English railway companies. In Britain scorn was heaped on the American products. One journalist, Charles Rous-Marten, who saw Baldwin products at the Paris Exhibition in 1900, compared them to the work of a village blacksmith! All this was a combination of wishful thinking and what would now be called spin-doctoring. The threat to the Glasgow

The second William Beardmore (1856-1936) became a partner in the famous Glasgow firm of iron producers and forgers in 1879, following the death of his father, and sole partner in 1887. Under his control the firm grew into a huge steelmaking, shipbuilding and engineering concern, but by the late 1920s it had run into serious financial difficulties. He became Lord Invernairn in 1921, in recognition of the contribution his firm had made to the production of armaments during World War I.

locomotive builders was real enough. Their response was to amalgamate. The union of Neilson Reid, Dübs and Sharp Stewart in 1903, to form the North British Locomotive Company, created the largest locomotive building company in Europe. There were over 7,500 employees and the three works had a combined capacity of 600 locomotives a year. This figure was never reached. In the best year, 1905, 573 engines were turned out and production fell away thereafter. It took World War I to bring a full order book again. In addition to locomotives much military equipment was made, including tanks, gun carriages, shells, sea mines and torpedo tubes.

By 1914 it should have been clear that there were problems ahead for the Scottish engineering industry. However, the effect of the World War I was to create a much increased demand for the traditional products and thus temporarily mask the problems. During the war the engineering industries in Scotland were fully stretched in their normal lines of business and in addition they had to undertake many things which for them were novelties. Among these were aircraft, aero-engines and tanks, which were built in large numbers. Following the end of the war in 1918 there was a boom period which led to a widespread belief that the good times were here to stay. There was to be a rude awakening within a few years and the depression which followed the boom was finally relieved only by the start of re-armament in the mid 1930s. It was becoming abundantly clear that the problems faced by the Scottish heavy industries were much more than another passing period of bad times. The heavy industries were unlikely ever to be able to contribute as much to the economy, or employ as many people, as they had done in the past. Many well-known firms had disappeared and there was considerable reorganisation among the survivors. Diversification became the watchword; new industries had to be found. In the Midlands and South of England the motor industry, and the numerous component industries which depended on it, had grown rapidly. The manufacture of domestic electrical equipment, such as cookers, vacuum cleaners, refrigerators and radio sets, was another growth industry. In the hope that such light engineering industries, along with other new industries, might be developed in Scotland, new 'industrial estates' were created. These offered small factories to rent and also space for the construction of larger factories to meet particular needs. The best known of the industrial estates set up before the start of World War II was that at Hillington, between Glasgow and Paisley. The ground was acquired in 1937 and after eighteen months there were sixty-seven tenants, some engineering-related, on the estate. Other industrial estates were established at Shieldhall (on ground belonging to the Clyde Navigation Trust), Dalmuir (on part of the site of the former Beardmore shipyard), together with smaller sites in Lanarkshire. The contribution of these industrial estates to the creation of employment was modest; only around 5,000 people worked in their factories in 1939. By then war was looming and the decision was taken to set up a factory at Hillington to build Rolls-Royce aero-engines. This became a huge operation. At its peak during the war, 25,000 people were employed and in total some 50,000 new and overhauled Merlin engines were turned out.

The story of World War II, as it affected the engineering industry, was much as it had been during the earlier conflict. Steel-making, heavy engineering, shipbuilding and repairing, were again fully stretched. In addition to the Rolls-Royce engine factory at Hillington, there was aircraft manufacture on a significant scale by the Blackburn Aircraft Company at Dumbarton, and an extensive operation, run by Scottish Aviation, was set up at Prestwick for the reception, conversion and modification of thousands of American-built aircraft which were flown across the Atlantic. The Manchester electrical engineering firm of Ferranti Ltd set up a factory in Edinburgh in 1943 to make gyroscopic gunsights for aircraft and later made radar and other electronic equipment.

The buoyancy created in the heavy industries immediately before and during the war again overshadowed their fundamental problems. Once more there was a post-war boom which this time lasted rather longer than that which followed World War I. After the war there was a determined attempt to maintain the aircraft-related industries set up during that period and the manufacture and overhaul of Rolls-Royce aero-engines continued and indeed flourished. The construction of industrial estates was resumed and new towns created, the first at East Kilbride. To occupy the factories, firms, many of them American, were encouraged to set up branch establishments making cash registers, office equipment, clocks and watches, domestic appliances and other goods. As well as light engineering products, there were a few American firms engaged in heavier engineering work, notably Euclid and Caterpillar, both making earth-moving equipment, and the diesel-engine maker Cummins. These industrial estates and new towns on greenfield sites were only made possible by twentieth-century technology in the form of the motor vehicle and a nationwide electricity supply. These removed some of the constraints which influenced the location of industry in the nineteenth century.

While the new towns grew and their factories brought new industries to Scotland, the old industries were facing a battle for survival, a battle which was all too often lost. The North British Locomotive Company was something of an icon. The immediate post-war years saw this firm, like many others, with a well-filled order book. Large numbers of locomotives, mainly steam, were built for both home and overseas railways. However, profits were hard to come by. Long-established customers were increasingly keen to nurture their own locomotive-building industries. Symptomatic of this trend was an Indian order placed in 1950 for one hundred WG class 2-8-2 locomotives. The next order was for boilers and other parts for another hundred locomotives to be assembled in India. Thereafter locomotives were wholly built in India in large numbers, without any outside help.

This period was the swan song of steam and customers increasingly required diesel and electric locomotives. In the building of a steam locomotive a high proportion of the work was done 'in house'. Building the new forms of motive power posed serious problems for all steam-locomotive builders. They could choose to collaborate with established makers of diesel engines and the electrical or hydraulic transmission equipment, and buy in these parts. Alternatively they could obtain manufacturing licenses and try to develop the necessary expertise themselves. The first route left them little for their own works to do, while the second needed new equipment and the retraining of the whole workforce into different ways of working. The diesel locomotive requires a higher degree of precision in its manufacture than a steam locomotive. NBL in fact used both approaches. They came to an agreement with the General Electric Company (a British company, not the American firm of the same name) on diesel and electric locomotives. In addition licenses were obtained for the manufacture of German diesel engines and hydraulic transmission systems from the firms of MAN and Voith respectively. For a mechanical engineering firm anxious to maintain its workload, hydraulic transmission was clearly more attractive than electric transmission. To assist with the changeover the firm took on several German engineers. Substantial orders were obtained from the British Transport Commission for both diesel-hydraulic and diesel-electric locomotives. Unfortunately, neither the re-equipment nor the re-training was adequate for the job in hand. The locomotives gave endless trouble and had very short working lives. Not surprisingly, there were no further orders from the British Transport Commission and significant orders from elsewhere proved impossible to obtain. As a result, the North British Locomotive Company went into liquidation in 1962.

Nigel Macmillan, in his book *Locomotive Apprentice* (1992), has written an illuminating account of his time with the company in the post-war years. He started his apprenticeship in 1947. In due course he went into the drawing office and then the project office. Here design proposals were prepared in response to enquiries from potential customers. From the vantage point of the project office it was clear that the firm was not winning the sizeable orders needed to ensure its future and he wisely decided to move on before the final collapse.

The North British Locomotive Company was by no means the only firm unable to make the transition from steam to diesel and electric traction. It was much easier for the makers of diesel engines and electrical equipment to turn themselves into locomotive builders. Until shortly before the failure there were around 7,500 people on the company's payroll. For Springburn in particular, where two of the three works were located, this was a major blow. Unfortunately there was more to come. When the railways of Britain were reorganised in 1923, the North British and the Caledonian Railways became parts of the London & North Eastern and London, Midland & Scottish Railways. Within a short time construction of new locomotives at the Cowlairs and St Rollox works in

Springburn had ceased, although both sites continued to handle a large amount of repair work. However, in 1968 Cowlairs was closed down and activity at St Rollox was much reduced. The Springburn district of Glasgow, known worldwide for its railway locomotives, has lost the industry which provided a living for most of its population for a hundred years. It was, too, an industry which had a strong family tradition, with several members of a family frequently in the same works and sons following fathers for more than one generation. There were to be many subsequent factory closures in Glasgow, but no district was harder hit than Springburn.

But the experiences of a few firms have shown that survival was possible. The Weir Group, incorporating Weir Pumps Ltd (well known for many years as G & J Weir), is now virtually the only large Scottish engineering firm which is truly world class in its field and still has its head office in Scotland. Marine work, which used to be a large part of the firm's output, is now only a small part of the business. Pumps are made for service all over the world in the oil industry, for power generation, water supply and general industrial use.

Although the North British Locomotive Company disappeared many years ago, railway equipment is still being built in Scotland. Andrew Barclay of Kilmarnock survived the nineteenth century despite many difficulties, including the tendency of the founder to indulge his passion for astronomy by building telescopes instead of running the business! In more recent times the firm has shown flexibility and a willingness to take up new ideas. The building of diesel locomotives was started in 1936 and both steam and diesel locomotives were made side by side as required by customers. In fact the last steam locomotive was not turned out until 1962. In their other traditional field, colliery winding-engines, a new type of brake was patented in 1951. By 1981, 132 sets of braking equipment had been installed. The design of control and braking systems for mine winding engines is a very specialised field. After the manufacture of steam winding engines ceased many Barclay brakes were fitted to existing steam engines as well as to new electrically powered winders. Barclay's also became involved in the manufacture of heavy shipyard machine tools. Despite the decline of the Scottish shipbuilding industry and therefore drastic contraction of the home market, one firm, Hugh Smith & Company of Glasgow, developed improved machines which were competitive throughout the world. From the 1960s they had so many

One of three trains, each consisting of a locomotive and two passenger coaches, built in 1999 for the Schneeberg mountain railway in Austria. Nearest the camera is the 730 hp (545 kw) diesel locomotive, equipped with rack and pinion drive. The locomotives were built by Hunslet-Barclay Ltd, while the passenger coaches were built in Austria using mechanical components supplied from Kilmarnock.

orders that some work was subcontracted to Andrew Barclay's. In 1977 Hugh Smith was taken over by one of the then fashionable conglomerates which decided to dispose of the company six years later. The business was acquired by Barclay's and since then manufacture of Hugh Smith machine tools has continued at Kilmarnock. With orders from the USSR, China, India, Egypt, Finland, Taiwan, Bulgaria, Italy, Egypt, Saudi Arabia and most recently the United States, the business is almost entirely export orientated. In recent years too, the amount of railway work at Kilmarnock has increased, and in addition to locomotives the firm is now involved in the construction and overhaul of railcars. In 1972 Andrew Barclay became linked to the Hunslet Engine Company of Leeds, which also manufactured industrial locomotives, becoming Hunslet-Barclay Ltd. Recently the Hunslet group became part of the Austrian firm Waagner-Biro. An interesting development following from this is the

Plate-bending machine by Hugh Smith, capable of bending steel plates up to 220 mm (8.6 inches) thick. Most machines built in recent years have been exported, but this one was for the oil rig builder Lewis Offshore, of Stornoway.
The inset shows a node or junction in the structure of a rig. The various cylindrical parts which were welded together to make up the node were shaped by a machine of this type.

89

manufacture at Kilmarnock of trains with rack-and-pinion drive for use on a mountain railway in Austria.

The work of Scottish engineers often took them abroad in the past. This still continues and at the time of writing engineers from Kilmarnock are in the United States erecting shipyard machinery and in Austria in connection with the trials of the trains for the mountain railway. Working abroad was not without its discomforts and hazards. In 1967-8 Andrew Barclay's despatched twenty-two diesel-hydraulic locomotives to East African Railways for service in Kenya and Uganda. These were followed by a further fifteen of a different design in 1972. After the military coup in Uganda which left Idi Amin in power, conditions there deteriorated to the point at which integrated operation of the railway system in the two countries was no longer possible. Unfortunately for the people of Uganda the main workshops and the stock of spare parts were in Kenya, and before long the Ugandan part of the system was at a complete standstill. After the removal of Amin in 1979, United Nations funding was made available to get the railway running again. The late Willie Rodie, who had been responsible for commissioning the locomotives when they were first delivered, was sent out to make an assessment of the work necessary to put those in Uganda back into running order. He later told the writer that on this visit he spent a week in what had been the best hotel in the capital, Kampala, and the only food available was bananas! Nevertheless, the job was done. He and other engineers from Kilmarnock returned on several occasions and, working under difficult conditions, carried out the necessary repairs.

There is clearly still a demand for at least some of the things for which the Scottish engineering industry was famous. With up-to-date products, closely tailored to the needs of the customer, more of the heavy industries could have survived. This did not happen and the emphasis remained on bringing new industries to Scotland. This could be done by the government making it difficult for firms to expand elsewhere, as in the case of the motor industry when the British Motor Corporation started production in 1961 of trucks and tractors at Bathgate and the Rootes Group began making the Hillman Imp car two years later at Linwood, near Paisley. Alternatively firms could be offered financial inducement to come to Scotland. The emphasis was shifting to electronics as American firms such as IBM and Hewlett-Packard and, increasingly, firms from

Japan, Taiwan and Korea, established branch factories. Many of these are simply production plants, with little involvement in engineering or commercial decision-making. There are notable exceptions, however. One is Hewlett-Packard, which came to South Queensferry in 1965, where it designs and manufactures test and measuring equipment for the telecommunications industry. Great things were expected of the North Sea oil industry as a source of engineering work for Scottish firms, but in fact very little was made in Scotland.

The motor manufacturers from the Midlands have packed their bags and gone home, as have many of the American firms who arrived in the years immediately after the end of World War II. The traditional Scottish engineering industry has continued to shrink to near vanishing point. There have been casualties also among the electronics firms. In 1997 the Taiwanese firm Lite-On opened a factory in Motherwell which was expected to employ 1,000 people making computer monitors. Already

The name Hewlett-Packard is popularly associated with computers and printers. However, their South Queensferry factory operates in a different field, making testing equipment, particularly for the telecommunications industry. The instrument shown is used for fault-finding in telephone exchanges. The associated software allows an engineer anywhere in the world to have access to the information which it collects.

they have decided to make these in China and the factory has closed. Indeed as a result of the economic problems of the Far East some firms, including the Korean giant Hyundai, have not got the length of occupying the new factories which were built for them.

The story of the Scottish engineering industry is a complex and fascinating one. Its study has not received the attention which its importance in shaping the present-day Scotland, for good and ill, merits. We have moved from being world class in the industries of the nineteenth century to a situation in the twentieth century of almost total dependence on technology from elsewhere. From the experience of several European countries, and indeed of a few firms in Scotland, the outcome could and should have been different.

7
The Engineers

The British engineering industry, and therefore the Scottish industry too, had the advantage of being first in the market, at home and overseas, with their revolutionary new products. By the end of the nineteenth century the initial advantages had largely disappeared and there was now strong competition from the engineering industries of recently industrialised countries, particularly the United States and Germany. In the course of the twentieth century it was to become increasingly clear that the British industry was ill-equipped to meet this competition. Nineteenth-century Scotland had been blessed with a plentiful supply of the basic raw materials, iron and coal, but these were wasting assets and by the end of the century iron ore was having to be imported. There was also plenty of low-cost labour as people left the Scottish Highlands and Ireland in search of the means of subsistence. The Scottish engineering industry also had the advantage of access to the markets of the greatest empire the world had ever known and to other markets with which there were long-standing commercial links. As the countries of the British Empire gained independence, not only were the markets opened up to manufacturers of other nations besides Britain, but there was a natural desire to manufacture things locally where possible. The wind of change became a gale and more foreign-made engineering products began to invade Britain, as well as the overseas markets which had been regarded as the special preserve of British exporters. There was, rather belatedly, consternation. Unfortunately most of the effort was directed towards the allocation of blame rather than attempting an analysis of the problem and suggesting solutions.

Politicians blamed the government or perhaps the previous government. Trades unions blamed incompetent managers, greedy owners or the machinations of international financiers. Managers and owners blamed a disputatious workforce, unwilling to accept new ways of working and prone to frequent strikes over both major and minor issues. The difficulty with all these 'explanations' lies in the timescale. They may indeed hold some element of truth, but the main reasons are more fundamental and have operated over a much longer time period than any of the allocators of blame can usually envisage. As already indicated, Scottish engineering firms were facing serious competition by 1900. Among all the explanations put forward for the difficulties faced by exporters, there has been a reluctance to suggest that in some areas Scottish (and for that matter British) engineering products were simply no longer good enough to compete effectively in world markets.

The Great Exhibition of 1851, staged at the Crystal Palace in London, was a confident celebration of British industrial superiority, although some cautionary notes were sounded even then. One of those involved in the organisation was Lyon Playfair, later Professor of Chemistry at Edinburgh University. Shortly after the Exhibition he expressed the opinion that Britain had to change its ways of doing things if it was not to be overtaken by European industry. Some of the comments on the 1862 International Exhibition in London indicate that there was increasing concern about possible foreign competition. What really set alarm bells ringing, however, was the Paris Exhibition of 1867. The range of high quality engineering products from European and American manufacturers indicated that Britain's superiority in the marketplace could be short-lived. Among those who took the trouble to investigate carefully there grew up a conviction that Britain's future competitors were devoting a much greater effort to education, and technical education in particular. Unless this was matched, the future for British manufacturing industry did not look good.

One result of this concern was a conference on technical education, held in Edinburgh in 1868 under the auspices of the Royal Scottish Society of Arts. This was attended by numerous academics, teachers, presidents of chambers of commerce, town councillors, engineers and industrialists from throughout Scotland. Among them was Sir David Baxter, the head of the large Dundee linen manufacturers Baxter Brothers & Company, who had recently provided an endowment to enable the University of Edinburgh to create its first engineering professorship. Also present was Professor Archer, Director of the Industrial Museum of Scotland set up in Edinburgh in 1854 and an ancestor of the present National Museums of Scotland. Many leading Scottish engineers attended, including John Scott Russell (a noted naval architect and shipbuilder who spent most of his career in England and built Brunel's monster ship *Great Eastern*), R W Thomson (a versatile engineer from Stonehaven who invented the

Almost hidden behind Neilson & Company's foremen and works manager in this group photograph is a small industrial locomotive. This was completed in 1862, soon after the firm's move from their original works in Hyde Park Street, in the Finnieston district of Glasgow, to their new Hyde Park works in Springburn.

pneumatic tyre and many other less well-known things) and Robert Douglas (the first British builder of the steam engine developed by the American engineer, George Corliss).

Resolutions were discussed and passed. It was declared that it was 'desirable and necessary that the principles of science should form an important element in the tuition of all classes of the community' and 'that for the more thorough instruction of all classes, it is desirable that Chairs of Applied Science be founded in the universities, and that branch Industrial Museums be established in the larger towns'. This serves as a reminder that the Industrial Museum of Scotland was originally created as an educational tool and showcase for the technology of the day, rather than a repository for historical artefacts. The resolution was moved, appropriately, by Baxter who said in his speech:

> … till the last few years, our country was supreme in the manufacturing industry of the world; but unfortunately we had been relying on that, and, while other countries had been educating their people, we had been left in the background, and subjected to very great competition from France, Switzerland, Belgium, not to say anything of America.

W J Macquorn Rankine (1820-1872), Professor of Civil Engineering and Mechanics at the University of Glasgow from 1855. Rankine, in his text books on steam engines, mechanics and shipbuilding was the first to present the scientific theories of the time in a form in which they could be used by engineers.

Scott Russell drew attention to the fact that in Zurich there were twenty professors of applied science, with salaries paid by the government and the buildings provided by the town. By that time similar institutions were already well-established across Europe. In 1794 France created the Ecole Polytechnique for the higher education of engineers, the earliest of its kind. In the 1820s the first of many Technische Hochschulen were established in Germany. These were technical universities and engineering research establishments. In the United States the Massachusetts Institute of Technology was founded in 1865. The first comparable establishment in Britain was Imperial College, London, set up early in the twentieth century.

In Scotland the Royal Technical College in Glasgow had grown out of the old Andersonian Institution. This was set up in 1796 with funds left by John Anderson, Professor of Natural Philosophy (Physics) at Glasgow University. He had started to give lectures for mechanics but in the process fell out with his colleagues. C A Oakley in *The Second City* described him as 'a remarkably brilliant but quarrelsome man'. His bequest, which turned out to be quite modest, was to be used to establish a second university. Fortunately the town council was sympathetic and offered his trustees accommodation for classes, initially in physics and chemistry. The royal accolade was received from King George V in 1912 and in the following year some classes at the Royal Technical College became recognised for degrees awarded by the University of Glasgow. In 1964 it became the University of Strathclyde. Two years later the Heriot Watt College in Edinburgh, whose origins went back to the School of Arts and Mechanics Institute of 1821, became Heriot-Watt University. The older universities had also set up engineering

G & J Weir's business was founded in the nineteenth century to supply such things as boiler feed pumps for use on board ships. Equipment for electricity generating stations now forms a major part of the ouput of Weir Pumps Ltd, and this photograph shows pumps being built in Glasgow for service in China.

departments, with the University of Glasgow appointing Lewis Gordon as its first Professor of Civil Engineering and Mechanics as early as 1840. He was followed in 1855 by the great W J Macquorn Rankine, author of important textbooks on applied mechanics, engineering thermodynamics and, in collaboration with others, shipbuilding and marine engineering. As already noted, Edinburgh University received the promise of an endowment in 1868 to enable it to do the same.

British technical education at all levels was to be the subject of many discussions, inquiries and reports in later years, but these did not result in the creation of a coherent strategy. Instead there was reluctant and piecemeal tinkering with the education system. Both in government and in industry itself there was enormous complacency. In 1937 E C Smith, a perceptive writer in the history of marine engineering and a marine engineer himself, wrote that the ships and their equipment constructed in Germany, France, Italy, the United States and Japan were fully equal to the best that Britain could produce. He could then continue confidently that our special talent for nautical affairs would enable our shipbuilders to match the achievements of their predecessors. This shows just how insidious the complacency born of long periods of undisputed superiority can be and how difficult it is even for those most closely involved to appreciate major changes at the time. Hindsight is of great assistance to subsequent commentators! Sixty years later there are still people who do not realise that things have changed. If complacency inhibits acceptance of changes within an existing industry, it is even more likely to discourage the acceptance and commercial development of completely new technologies. Two of the most important of these in the late nineteenth century, the internal combustion engine in its various forms, and electrical engineering, have been discussed in Chapter 5. In Scotland, and in Britain as a whole, while there was much complacency, both new industries were also hindered by a shortage of engineers with the necessary education and training.

By the early years of the twentieth century the best university-level technical education provided in Britain was of a high standard, and probably comparable with what was available anywhere else, but the number of students was a fraction of those in the main competitor nations. Overall, technical education for all levels of the British workforce remained decades behind that available in Europe. Alan Dunbar's account, in *Fifty Years with Scottish Steam*, of his own apprenticeship with the Caledonian Railway as a locomotive fitter, was mentioned earlier. There was no formal training within the company and little incentive to seek enlightenment elsewhere. He does note, however, that there were some apprentices who attended classes at the 'Tech' [Royal Technical College] several nights a week, in addition to putting in a full working day. There were also a number on sandwich courses, attending the university for six months of the year and working at St Rollox for the other six months. Both groups finished their apprenticeships with a spell in the drawing office. Alan Dunbar had to rely on his own initiative, and on the help of some of the charge hands

Machining shells in the 'Mons' factory at the works of the North British Locomotive Company, Glasgow. Production started late in 1916 and 864,551 shells of various sizes were produced. Women made up more than two-thirds of the workforce which varied from about 1,000 to 1,500, but the foremen and tool setters were all men.

and journeymen, to enable him to overcome some of the deficiencies of the system. However, these deficiencies were all too real and widespread within the engineering industry. It did not help that there was a reluctance in industry to believe that training really mattered, and an unwillingness to link the long term effects of deficiencies in training to the problems increasingly encountered in world markets.

It was suggested above that British products have often been uncompetitive in the market place simply because they were not good enough. In other words, they were not as well designed as those of competitors. 'Design' is not something newly invented in the USA in 1930s. It is and always has been the core activity of engineers. In the years after World War II much was heard about the high standard of design of Danish and Swedish furniture, glassware and ceramics, and the failure of Britain to produce goods of comparable quality. This was a fair comment and a matter for regret. What was much more important in its long-term effect, but less widely discussed, was that Britain was unable to match the Scandinavian (and other) countries in the design and manufacture of engineering products.

Further Reading

BRITISH ASSOCIATION FOR THE ADVANCEMENT OF SCIENCE: *Some of the Leading Industries of the Clyde Valley* (Glasgow, 1876).

BRITISH ASSOCIATION FOR THE ADVANCEMENT OF SCIENCE: *Local Industries of Glasgow and the West of Scotland* (Glasgow, 1901).

BROWNLIE, J S: *Railway Steam Cranes* (Glasgow, 1973).
Deals with the development and operation of steam cranes and the history of the makers, including the various Scottish firms.

DUNBAR, Alan G and I A GLEN: *Fifty Years with Scottish Steam* (Newton Abbot, nd).

GRIFFITHS, Denis: *Steam at Sea, Two Centuries of Steam-Powered Ships* (London, 1997).
The scope of the book is world-wide, but there is much about the work of the Scottish marine engineers.

HARVEY, Robert: *Early Days of Engineering in Glasgow* (Old Glasgow Club Transactions, 1919).

HARVEY, W S and G DOWNS-ROSE: *William Symington, Inventor and Engine Builder* (London, 1980).

HUGHSON, Martin: *John Robertson, Engineer* (Barrhead & Neilston Historical Association, 1989).

HUME, John R and Michael S MOSS: *Beardmore, the History of a Scottish Industrial Giant* (London, 1979).

HUME, John R: *Scotland's Industrial Past* (Edinburgh, 1990).
Includes a summary of preserved material in Scotland.

MACMILLAN, Nigel S C: *Locomotive Apprentice at the North British Locomotive Company* (Brighton, 1992).

MOSS, Michael S and John R HUME: *Workshop of the British Empire* (London, 1977).

NEWMAN, Brian: *Plate and Section Working Machinery in British Shipbuilding 1850-1945* (Centre for Business History in Scotland, University of Glasgow, 1993).
One of the few books dealing with the development of shipyard machine tools, in which Scottish firms played an important part.

NICOLSON, Murdoch and Mark O'NEILL: *Glasgow, Locomotive Builder to the World* (Glasgow, 1987).

NICOLSON, Murdoch: *Glasgow, Locomotive Builder to Britain* (Glasgow, 1998).

OAKLEY, C A: *Scottish Industry Today* (Scottish Development Council, 1937).

OAKLEY, C A: *Scottish Industry* (Scottish Council (Development and Industry) 1953).
Both of Oakley's books contain useful information on the state of the engineering industries when they were written, together with some historical notes. They have a determinedly optimistic tone, however, and with the benefit of hindsight it is clear that this optimism was misplaced.

TURTON, Alison and Michael MOSS: *The Bitter with the Sweet, The History of Fletcher and Stewart, 1838-1988* (Derby, 1988).
The histories of two firms of sugar machinery makers, Duncan Stewart & Company of Glasgow, and the Derby firm, George Fletcher & Company. The companies came under common ownership in 1958.

WEAR, Russell: *Barclay 150, 1840-1990* (Kilmarnock, 1990).
A history of the famous Kilmarnock engineering firm, Andrew Barclay, Sons & Company Ltd.

Places to Visit

Anstruther   The Scottish Fisheries Museum has examples of Scottish-built engines used in fishing vessels.

Blairlogie   Keithbank Mill is a former flax and jute spinning mill, now open to visitors, which houses a large water wheel and a single-cylinder horizontal steam engine, both made in Dundee about 1865.

Coatbridge   Summerlee Heritage Park has an important collection of machine tools and other engineering items.

Dumbarton   Displayed outside the building housing the Denny Ship Model Experiment Tank of 1883 (now part of the Scottish Maritime Museum) is Robert Napier's first marine engine, built in 1823 for the paddle steamer *Leven*.

Dundee   The McManus Galleries contains material relating to Dundee industry, including engineering.

At the Verdant Works, a former jute mill, the Dundee Heritage Trust has a museum of the jute industry, containing examples of the types of machinery used. The various processes are demonstrated regularly. Much of the machinery was made locally.

Edinburgh   The Museum of Scotland includes engineering exhibits. More engineering material, including a collection of large stationary steam engines, is held at the Granton Research Centre of the National Museums of Scotland.

Garlogie, Aberdeenshire

The Garlogie Power House Museum has been created round a mid nineteenth-century beam engine which powered a woollen mill. The engine has survived in its original house although the mill which it drove has long gone.

Glasgow   The Museum of Transport has Glasgow-built locomotives and marine engines, both model and full size.

Greenock   The McLean Museum has material relating to local industry including engineering.

Newtongrange, Midlothian

The Scottish Mining Museum is at the former Lady Victoria colliery. The original steam winding engine built in 1894 by Grant, Ritchie & Company of Kilmarnock is still in situ. There is an important collection of underground machinery and other equipment used in the industry.